Electronics Projects
using Electronics Workbench

M. P. Horsey

 Newnes

Newnes
An imprint of Butterworth-Heinemann
Linacre House, Jordan Hill, Oxford OX2 8DP
A division of Reed Educational and Professional Publishing Ltd

A member of the Reed Elsevier plc group

OXFORD   BOSTON   JOHANNESBURG
MELBOURNE   NEW DELHI   SINGAPORE

First published 1998

© M. P. Horsey 1998

British Library Cataloguing in Publication Data
A catalogue record for this book is available from the British Library

ISBN 0 7506 3137 6

Library of Congress Cataloguing in Publication Data
A catalogue record for this book is available from the Library of Congress

PLANT A TREE

British Trust for
Conservation Volunteers

FOR EVERY TITLE THAT WE PUBLISH, BUTTERWORTH-HEINEMANN
WILL PAY FOR BTCV TO PLANT AND CARE FOR A TREE.

Composition by Genesis Typesetting, Rochester
Printed and bound in Great Britain by
Biddles Ltd, Guildford and King's Lynn

Contents

Points to watch

Listed below are the features that may seem odd when first using Electronics Workbench.

1 Electronics Workbench can calculate voltages around your circuit much more quickly and accurately if you use the 'ground' symbol to fix a particular part of your circuit to 0 V. Hence, in all the circuits included in this book, a 'ground' symbol is used even when not essential.

2 The 'activate' switch at the top right-hand corner of the screen should not be used as an on/off switch. Think of it rather as a 'run/pause' switch. For example, if you switch on a bulb, then turn off the 'activate' switch, the bulb will remain lit. The program is showing you the state of the circuit at the moment *before* the 'activate' switch was turned off.

3 Keep an eye on the time counter shown at the right-hand side of the screen. If the counter freezes, then Workbench is having to perform a large number of calculations – be patient! Note that a flashing bulb, for example, may appear to flash at the wrong rate as it is likely that the simulation is running much more slowly than real time. However, the rate in relation to the counter time should be correct. If Workbench decides that nothing else can happen in your circuit the words 'steady state' appear and the simulation pauses. This is irritating and can be avoided as follows: Include a switch from the parts box in series with your battery. This will prevent the software entering the 'steady state' mode.

4 The software assumes that wires and switches are perfect conductors. Hence, two switches in *parallel* will cause problems. If switches *have* to be wired in parallel, then place a 1 Ω resistor in series with each parallel switch.

5 Bulbs are rated at a certain power (in watts) and voltage (in volts). The voltage across a real bulb can, in practice, be much higher than its rated voltage. However, in Electronics Workbench the voltage stated is the *maximum*, beyond which the bulb burns out.

6 The potentiometer (which can also act as a variable resistor) is calibrated in terms of percentage. For example, a 1 kΩ potentiometer set at 50 % means that the wiper is halfway along the resistor. This percentage is reduced by pressing R, or increased by pressing SHIFT R. Note that if the caps lock button is pressed, the percentage cannot be decreased.

The author would like to thank the editor of *Everyday Practical Electronics* magazine for permission to use material originally published by the author in the magazine.

Chapter 1

Some basic concepts

Simple circuit

The best way to test a new system is to begin with a circuit that has a known result, so begin with the simplest possible circuit!

1 Connect up the circuit shown in Figure 1.1(a).

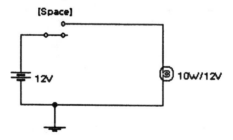

Figure 1.1a

2 Activate the circuit by clicking on the activate switch.
3 Use the space-bar to toggle the switch on and off.
4 Double-click the bulb and change its voltage to 10 V. Test the circuit.
5 Double-click the battery and change its voltage to 6 V. Test the circuit.
6 Now change the battery voltage to 4 V.

Conclusions

If the stated voltage of the bulb is exceeded, the bulb blows. The bulb will not light if the voltage across it is less than half its stated voltage.

Taking measurements

1 Place a voltmeter in parallel with the bulb and an ammeter in series with it, as shown in Figure 1.1(b). **Note**: If you drop the ammeter into a wire it will connect automatically. If necessary, *rotate* its connections using COMMAND R. The voltmeter should confirm the battery voltage.

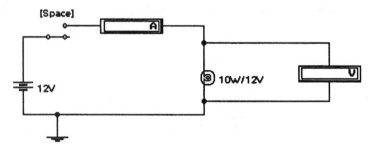

Figure 1.1b

2 Change the battery voltage to 10 V and the bulb ratings to 10 W, 10 V; the ammeter should read 1 A. This confirms the formula:

$$\text{Power} = \text{voltage} \times \text{current}$$

Bulbs in series

1 Try out the circuit shown in Figure 1.2. Note that the battery has been increased to 24 V. What do you notice about the readings on the voltmeters?

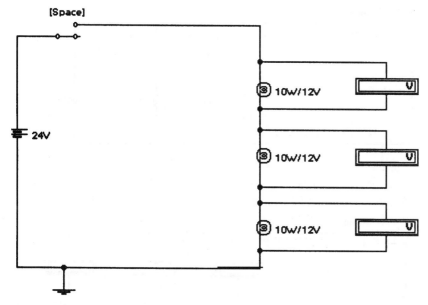

Figure 1.2

2 Double-click on any bulb and change its power rating (in watts). The voltage across the bulbs will now be different, but if you add together the voltage across each bulb, the total should still equal the battery voltage.

Conclusion

When bulbs (or resistors) are connected in series, the individual voltages across each bulb add up to the total voltage across the chain.

Bulbs in parallel

1 Try out the circuit shown in Figure 1.3(a). Check that the battery is set to 12 V, and use new bulbs from the parts bin.
2 Add ammeters as shown in Figure 1.3(b).

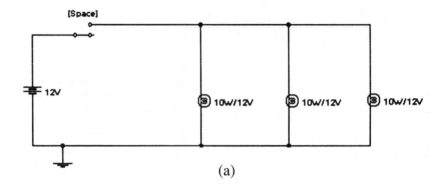

(a)

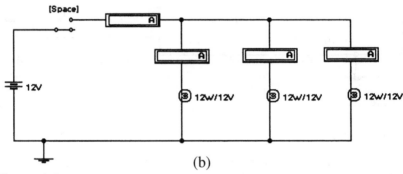

(b)

Figure 1.3

3 Change the power rating (in watts) of each bulb to 12 W. This
  will ensure that each bulb passes 1 amp (1 A).

Conclusion

The sum of the current used by each bulb equals the total current
supplied by the battery.

Voltage, current and resistance

1 Try out the circuit shown in Figure 1.4.
2 Write down the readings on the voltmeter and ammeter.
3 Now divide the voltmeter reading by the ammeter reading. What
  do you notice about the value of the resistor?

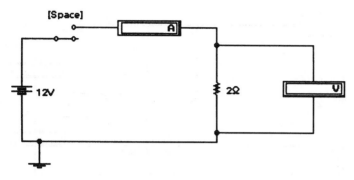

Figure 1.4

4 Change the value of the resistor to 4 Ω.
5 Switch on the circuit. Does the resistance still equal voltage/current?
6 Try other values for the resistor.

Conclusion

The relationship between voltage, current and resistance is known as Ohm's Law. This shows the following relationship:

$$\text{Resistance} = \text{voltage/current}$$

Potential dividers

'Potential' is another word for voltage. A potential divider is simply a pair of resistors. This type of circuit is used to set a voltage in a circuit. It is probably the most widely used technique in electronics.

For convenience when calculating the result, the bottom resistor is fixed to 0 V as shown in Figure 1.5. The input voltage is fed to the top resistor, and the output voltage is obtained from the junction between the resistors.

1 Try out the circuit in Figure 1.5.
2 Notice how the two *equal* resistors divide the voltage *equally*.
3 Change the lower resistor to 2 kΩ. Notice that the voltage across the 2 kΩ resistor is twice the voltage across the 1 kΩ resistor.

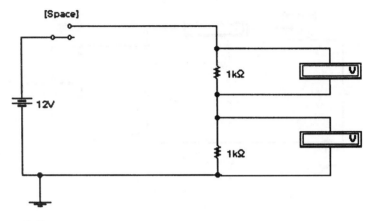

Figure 1.5

4 Make both resistors 2 kΩ. Note the readings. If both resistors
  have the same value, will they always divide the voltage
  equally?
5 Try other values to find out.

Practical circuits

The potential divider is used to reduce a voltage by a fixed ratio. It
will be employed many times in the projects in this book. It is also
used as part of an input or control circuit. For example, in a light-
sensing circuit, a light-dependent resistor (LDR) may be used in
series with a variable resistor. The output voltage from the junction
will rise and fall as the light changes.

Daylight-sensing circuit

1 Try out the circuit in Figure 1.6.
2 Double-click the lower variable resistor, and change the key to L.
3 Check the operation of the variable resistors: whenever R is
  pressed, the upper variable resistor is reduced in value by 5%.
  Whenever L is pressed, the lower variable resistor is reduced by
  5%. To increase the value of either variable resistor, hold down
  the SHIFT key while pressing R or L. **Note**: if the caps-lock button
  is set, the percentage cannot be reduced.

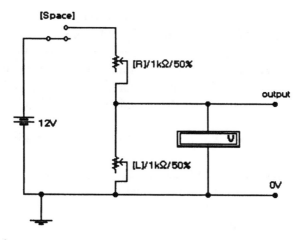

Figure 1.6

The lower variable resistor represents the LDR. Pressing L decreases its resistance and simulates increasing daylight. Pressing SHIFT L increases its resistance and simulates the falling daylight.

The upper variable resistor provides 'user control'. It determines the point at which the circuit responds to changing daylight.

A simple, if crude, way of using the circuit to make an automatic lamp is shown in Figure 1.7. Add the extra components as shown. Electronics Workbench assumes a perfect relay coil with a DC resistance of virtually zero. This will give odd results in this

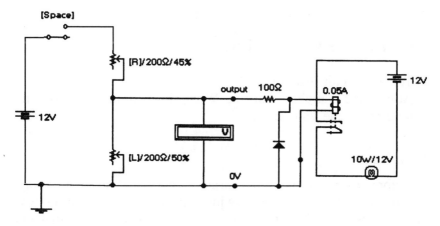

Figure 1.7

application, hence the addition of the $100\,\Omega$ resistor shown. In a practical circuit, the $100\,\Omega$ resistor may be omitted.

The diode cures a problem caused by 'back emf'. Whenever the relay is switched on or off, high voltages are generated in the coil. These can cause a problem particularly if a transistor is used to drive the relay. This will be discussed further in Chapter 2.

1 Double-click both variable resistors and set their values to $200\,\Omega$.
2 Simulate decreasing daylight by pressing SHIFT L. If necessary, reduce the value of the upper resistor by pressing R. It should be possible to make the lamp switch on and off.
3 Change the extra battery voltage to 230 V, and the lamp to 230 V. This simulates the control of a real mains lamp.

Shortcomings

This circuit is crude, as a large current is required to operate the relay, and this current upsets the output voltage produced by the potential divider. Have you noticed that the voltmeter readings are lower than expected? Try making both variable resistors equal. Does the voltmeter read 6 V?

A well-designed circuit will require an output current of less than a tenth of the current flowing through the potential divider chain. In Chapter 3, a transistor is employed to make the circuit work much more reliably.

Odd results with resistors

1 Try the circuit in Figure 1.8.
2 Adjust the variable resistor. What do you notice about the reading on the voltmeter?

Conclusion

Resistors, whether fixed or variable, will not cause a voltage difference unless current is flowing. Since the current through the voltmeter is very small, the value of the resistor appears to make no difference to the voltmeter reading.

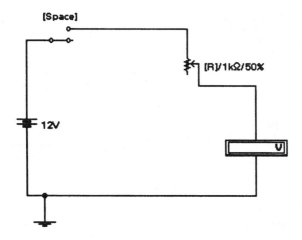

Figure 1.8

Voltmeters

Double click the voltmeter in Figure 1.8, and change its resistance to 1 kΩ. (Voltmeters have resistance and can affect the circuit being tested.) You have now created the worst voltmeter ever used! It uses so much current that it has a large effect on the circuit. It is rather like a tyre pressure gauge that lets out so much air from your tyre that the pressure is reduced whenever the gauge is used. Change the voltmeter back to 1 MΩ resistance. This is a typical value for most modern voltmeters.

A lamp dimmer

1 Add a lamp to your circuit as shown in Figure 1.9. The current flowing through the lamp will cause a voltage drop across the variable resistor. In fact, the current required by the lamp (nearly 1 A) is much too large for the 1 kΩ variable resistor.
2 Double-click the variable resistor and reduce its value to 30 Ω. In other words, its maximum value is 30 Ω and, when set to 50 %, its value is 15 Ω.

It should now be possible to control the voltage across the lamp in regular steps between 0 V and 12 V. The software will not simulate

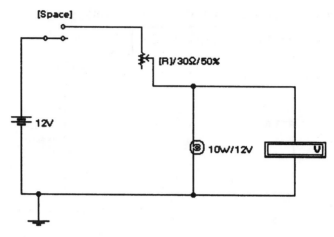

Figure 1.9

a lamp changing in brightness, but the reading on the voltmeter provides a guide.

In the past, this type of circuit was commonly used in the theatre for light-dimming systems. However, it has considerable draw-backs: a large current flows through the variable resistor, making it very hot. Modern light-dimmers employ some clever electronics to control lamp brightness.

Questions

Full written answers, complete with Workbench circuits, are available on the accompanying disk. See p. 224 for details.

1 A set of Christmas tree lights bought in the USA had a label stating 'For use on 120 V'. Each bulb had a rating of 12 V, and they were all wired in a series chain.

(a) How many bulbs were in the set?

(b) Draw a circuit diagram to show how the bulbs were connected. Employ a 120 V battery, and 12 V bulbs. Include an ammeter in series to check the current.

(c) When the bulbs were examined more carefully, each one was found to be rated at 12 V 3 W. How much current was used from the mains supply?

2 A 12 V car battery is used to supply the following:

- two headlamps each rated at 12 V 48 W
- two rear lamps each rated at 12 V 12 W
- a single reversing lamp rated at 12 V 24 W.

(a) Draw a circuit diagram showing how all these items are connected, with a switch operated by L controlling all the headlamps and rear lamps and a switch operated by R controlling the reversing lamp.
(b) What is the total power rating (in watts)?
(c) How much current is used by each headlamp?
(d) What is the total current supplied by the battery when all the lights are on?

3 A resistor of value 5 Ω is connected across a 20 V battery.

(a) State the voltage across the resistor.
(b) Calculate the current through the resistor.

4 Two resistors, each of 6 Ω are wired in series and connected across a 12 V battery.

(a) State the voltage across each resistor.
(b) Calculate the current flowing from the battery.

Chapter 2

Projects with switches, LEDS, relays and diodes

Wiring up a doll's house

1 Try the circuit shown in Figure 2.1.

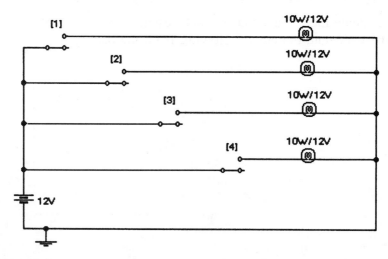

Figure 2.1

2 Double-click on each switch, and set each key to the appropriate number as shown. Now, instead of pressing the space-bar to turn a switch on or off, the numbers 1, 2, 3 or 4 must be pressed instead.

Notice how each lamp can be independently controlled by a particular switch. Each switch/bulb circuit is connected in *parallel* across the power supply.

Further work

Most houses have at least one lamp (the landing light, for example) controlled by two switches.

1 Select another lamp and two switches from the parts bin.
2 Double-click each switch and set the key for one to U (for upstairs) and for the other to D (for downstairs).
3 Now connect up the switches and lamp so that either switch can turn the lamp on, and either switch can turn it off. **Note:** so far, we have used single pole, single throw (SPST) switches for simple on/off switching. This circuit requires single pole, *double* throw (SPDT) switches, often called 'two-way' switches. In fact, the switches in Workbench are all SPDT types, so it is just a matter of using both 'ways' to obtain the solution.
4 The two-way circuit is provided in Figure 2.2. Try to incorporate it into Figure 2.1.

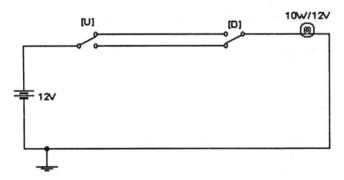

Figure 2.2

Train signal controller (3-wire)

Try the circuit shown in Figure 2.3. **Note:** the correct LED might not light up immediately the switch is changed; there may be a pause – it depends upon the speed of your computer.

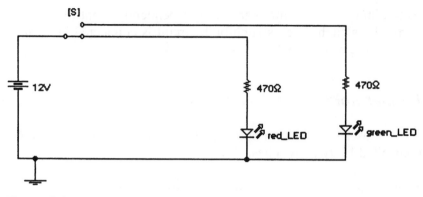

Figure 2.3

LEDs or bulbs may be used in this circuit. However, LEDs require resistors as shown to limit the current to a safe level. An LED connected without a resistor may appear to work, but will soon burn out and/or destroy the power supply. If bulbs are used in place of LEDs, then the resistors should not be included.

Selecting LED resistor value

Omit this section if you prefer to avoid mathematics!

Workbench assumes that 20 mA is required by an LED. In real life, 10 mA will make a good quality LED light up satisfactorily, and special low-current LEDs are also available, which require only 2 mA.

An LED may be powered from any supply above 3 V; a higher voltage simply requires a higher resistor value.

To calculate the value of the resistor required, first decide how much current you wish to flow through the LED. We will assume 20 mA. Then establish the forward voltage drop (V-drop) expected across the LED. This figure should be stated in the LED specifications. The figure is around 1.8 V for red LEDs and 2.1 V for green and yellow. We will assume 2 V for all.

The formula required is:

$$R = (\text{Supply voltage} - \text{LED V-drop})/\text{current required}$$

in our case:

$$R = (12 - 2)/0.02 = 500\,\Omega$$

Hence, our choice of $470\,\Omega$ provides the nearest resistor value available that will allow at least $20\,mA$ to flow.

Train signal controller (2-wire)

There may be a considerable distance between your switch and the signal. Using a pair of wires, rather than three, may be more cost effective. However, we now need a more complicated switch.

Try the circuit shown in Figure 2.4. The fact that an LED is a *diode* is put to good effect. Diodes conduct electricity in only one direction. So, if the LEDs are connected 'in reverse parallel', as shown in Figure 2.4, then only one will light. If the polarity of the supply is reversed, then the other LED will light.

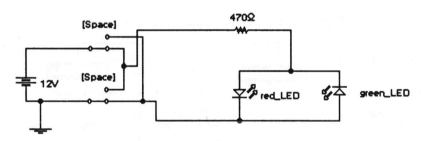

Figure 2.4

A DPDT switch is required to reverse the polarity. This is achieved by selecting two switches from the components window, and setting them both to the *same* key. When this key is pressed, they will both change-over. Follow the connections carefully to see how the polarity of the supply is reversed whenever the switches are operated.

Practical note

It is possible to buy 2-colour LEDs if preferred. These look like a normal LED, but inside there is a red LED and a green LED wired in 'inverse parallel'. When the switch is changed the LED appears to change from red to green and vice-versa.

Further work

1 Replace the LEDs with bulbs. The circuit will no longer work properly as bulbs conduct equally in both directions.
2 Add the necessary components to the circuit to make it work properly.

Working with diodes

Fig 2.5 shows a circuit designed to explore the diode. Activate the circuit and switch on by pressing the space-bar.

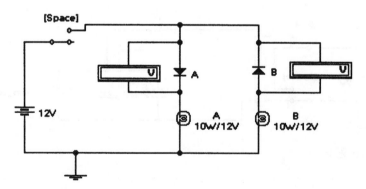

Figure 2.5

Note that only bulb A lights up. If you think of the diode symbol as an arrow, it points in the direction of the conventional flow of current. (Conventional flow means that the current flows from positive to negative.)

The voltmeters give the full story. Diode B faces towards positive and so no electricity can flow. Voltmeter B reads 12 V, showing that the entire battery voltage is across the diode, hence there is no voltage across the bulb. *Place a voltmeter in parallel with the bulb to prove this point, if necessary.* Diodes cannot withstand an unlimited voltage. The maximum voltage that can be tolerated is known as the peak inverse voltage (PIV). For example, diode type 1N4001 has a PIV of 50 V. Type 1N4004 has a PIV of 400 V (helpful for mains use): type 1N4148 has a PIV of 75 V.

Diode A allows current to flow from positive towards the bulb. In theory, the voltage across it would be zero, and the full 12 V supply would be across the bulb. But diodes are not perfect conductors; nor are they ordinary resistors – they do *not*, in fact, obey Ohm's Law. There is a voltage difference across the diode (known as the Voltage drop, or V-drop), which ranges from about 0.6 V to 1 V for silicon diodes (less generally, for germanium diodes). The odd – but very useful feature – is that this voltage changes by only a small amount if the current flowing is changed.

1 Try changing the power of bulb A to 100 W so that 10 times the current flows through the diode. The V-drop across the diode, which was previously about 0.8 V (800 mV), hardly changes.
2 Try reducing the power to 1 W. Again, the V-drop reading only changes by a very small amount.

Practical points

When selecting diodes begin by choosing a diode capable of carrying the required current. For example, type 1N4148 (very small and cheap) can conduct 100 mA. Type 1N4001 (and 1N4004) can conduct 1 A. Also check that your diode has a sufficiently high PIV.

The ends of the diode are generally known as anode and cathode. The anode is the arrow head in the symbol; it points towards the cathode. In a real diode the cathode is generally denoted by a black or silver band.

Motor reversing circuit

A motor may be connected to the DPDT switch as shown in Figure 2.6. A second switch is used to switch the motor on and off. Ensure that it is operated by a different key. The motor is simulated with a voltmeter. A positive reading indicates 'forward'; a negative reading indicates 'reverse'.

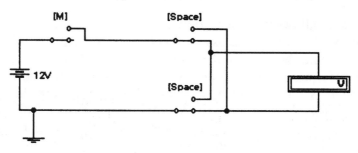

Figure 2.6

Relay circuit

You may have already discovered that the Electronics Workbench relay operates rather slowly (unless you have a super-fast computer). This is because Workbench is making a number of calculations based on the fact that the relay coil is an *inductor*. In other words, the coil produces a magnetic field and, whenever this field rises or falls in magnitude, it generates a voltage (emf is a more accurate term than voltage in this context), which in turn affects the current flowing through the coil. If all this sounds complicated, it is, and that is why Workbench has to perform so many calculations when the relay is turned on or off.

The inductive effect need not concern us in simple relay circuits, and so the voltage-controlled switch is employed as shown in Figure 2.7(a). This operates just like a relay, but we ignore any inductive effects, and so Workbench runs much more quickly.

When selecting the voltage controlled switch, note that both the 'turn-on' and the 'turn-off' voltages can be set. The turn-on voltage can remain at the default setting of 1 V, but the turn-off voltage should be increased to 0.5 V.

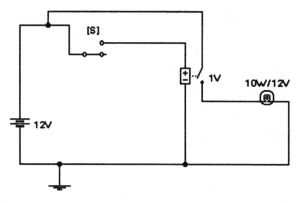

Figure 2.7a

Try the circuit. The voltage-controlled switch (which, in real life, would be a standard relay) may appear pointless in this circuit, as the lamp is only being switched on and off; something that could be done with a simple switch. However, relays are very useful devices. For example, Figure 2.7(b) shows how a relay can control a different and independent voltage. A safe supply of 12 V can be used on the 'primary' side, but the relay contacts can control a much higher voltage.

Relays are often employed to control mains supplies. In Figure 2.7(b), a supply of 120 V AC is controlled. Note that 120 V is the default AC supply. This is because Electronics Workbench was designed in Canada, and both Canada and the USA employ mains supplies of 120 V. In the UK and throughout Europe, the mains supply is 230 V.

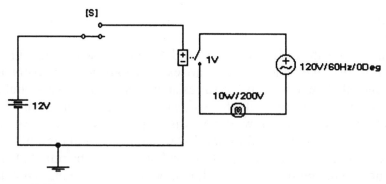

Figure 2.7b

You may have noticed that the lamp voltage is much higher than the AC supply. Workbench causes a lamp to blow if the voltage across it rises above its rated voltage. As an AC supply of 120 V actually 'peaks' at 170 V, the bulb must be set above this figure. We will discuss AC peaks in Chapter 4.

Latching relay circuit

A latching circuit, as the name suggests, remains *on* when triggered. There are many ways of creating latching systems, and, later in the book, we will examine a more elegant latching circuit based on a logic gate bistable.

However, a simple relay can be made to latch quite easily. The required circuit is based on Figure 2.7(a), with the addition of a 'feedback loop'. The resulting circuit is shown in Figure 2.8.

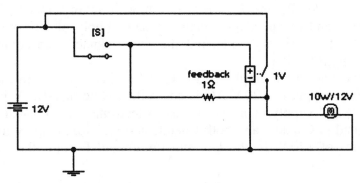

Figure 2.8

In real life, the feedback loop would be a normal wire link, with *no* resistor. However, another quirk arises in Workbench. The software assumes that switches have zero resistance, and so, if two switches are connected in parallel, the software will report an error. The solution is to place a low-value resistor in series with one of the switches – hence the need for the 1 Ω resistor shown in Figure 2.8.

Note that, when the switch is toggled on and off, the relay latches on, and so the bulb remains on. It can be reset by clicking the main run/pause switch.

A more elegant way of resetting the circuit is to place another switch in series with the resistor in the feedback loop. Double click this switch to set key R (for reset). Assuming that the first switch is operated with key S (set) the circuit can be made to set/reset as required.

A circuit that can be made to latch and unlatch can store one *bit* of information. This is the principle of electronic memory. Eight circuits like this (i.e. 8 bits, or 1 *byte*) could be used to store a letter of the alphabet, or a number.

Alarm system

A simple alarm system can be based around a relay. Begin with the circuit shown in Figure 2.8, but replace the bulb with a buzzer to represent the siren.

The switch (S) represents a microswitch, which is triggered when a door is opened. The reset switch (R) would normally be key-operated to reset the alarm after it has been triggered. Figure 2.9 shows the reset switch in series with the power supply. This is to enable the user to switch off and reset the alarm in a single operation.

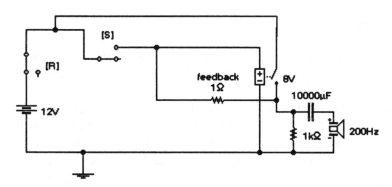

Figure 2.9

House alarm systems can cause immense irritation to neighbours, and they should never be allowed to sound for more than 15 minutes. A capacitor in series with the buzzer provides a crude method of time control. The capacitor will appear to conduct electricity until fully charged; therefore the buzzer sounds during the charging process. Larger capacitors take longer to charge. Try out different capacitor values by double-clicking the capacitor. If the times appear odd, check the time display on the right hand side of the screen – the computer may be running much more slowly than real time.

The charging time also depends upon the current flowing and, in this circuit, the current is determined by the buzzer. Double-click the buzzer and change its current setting.

For reliable results, double-click the voltage-controlled switch ('relay') and change its 'on voltage' to 8 V and its 'off voltage' to 7 V. This will ensure that the charge on the capacitor does not keep the relay latched on when the reset switch is turned off.

A resistor is required to discharge the capacitor, ready for the next time the circuit is triggered. Try changing its value to, say, 1 MΩ. The alarm circuit will work the first time, but subsequent triggering may not cause the buzzer to sound because the capacitor will not have had time to discharge.

In practice, a loud siren will require more than 100 mA (0.1 A). A capacitor in excess of 1000 µF will be physically large and expensive.

Questions

Full written answers, complete with Workbench circuits, are available on the accompanying disk. See p. 224 for details.

1 Draw the circuit diagram of the lighting system for a Doll's House. Include a single 12 V battery, two down-stairs room lights with individual switches A & B, one upstairs room light with a switch C, and a landing light, which can be operated with upstairs and downstairs switches D & E.

2 Show the circuit diagram of an LED operating from a 6 V battery. Include a series resistor. Calculate the value of the resistor required such that 25 mA flows through the LED. Check your calculations by adding meters to your circuit.

3 Draw a circuit diagram to show a 12 V battery driving 2 LEDs in series. Calculate the series resistor needed if the current required by each LED is 20 mA.

4 Draw a 12 V battery on the left hand side of the screen, and two 12 V bulbs on the right. Label the bulbs R for red and G for green. Add a suitable switch next to the battery to make a 'signal controller'. The controller must be able to light either the red bulb, or the green bulb only. This can be achieved with only *two wires* linking the controller to the bulbs by adding diodes to the circuit.

5 Draw a circuit diagram consisting of 3 bulbs (rated at 12 V), 3 switches, a 10 V battery and any necessary diodes. Arrange it so that:

(a) Switch 1 makes *only* bulb 1 light up
(b) Switch 2 makes *only* bulbs 1 & 2 light up
(c) Switch 3 makes all three bulbs light up.

6 Design an alarm system with a set switch (operated by pressing S), that acts as the trigger switch and sounds the alarm, a reset switch (operated by pressing R), a relay that latches when triggered, and a buzzer.

Chapter 3

Transistors

Transistor switch

The circuit shown in Figure 3.1 is not a complete project, but a very useful module, which allows a tiny amount of current to switch on a large current. The transistor can be likened to a relay in this respect, but it is far more flexible, less expensive and smaller. Its only limitation is that it cannot handle high voltages.

In Figure 3.1 the bulb remains off because nothing is connected to the transistor input. Try putting a wire link between the input dot and positive. The bulb should now switch on. If not, check that you

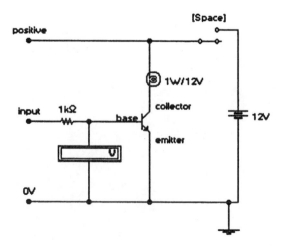

Figure 3.1

have activated the circuit by clicking on the run/pause switch, and have pressed the space-bar to switch on the circuit.

The three connections of the transistor are known as its base, collector and emitter. A voltmeter is included to check the voltage between the base and emitter. If this voltage is less than about 0.5 V, the transistor is turned off. If it rises beyond 0.7 V, the transistor turns on. The voltage across the base/emitter is unlikely to rise above 0.9 V. The base/emitter junction is like a forward-biased diode.

The resistor in series with the base is needed to ensure that excessive current is not allowed to flow into the base of the transistor. If the base/emitter voltage is forced to rise beyond 0.9 V due to excessive current at the base, the transistor is likely to burn out.

Later, we will see how the value of the base resistor may be calculated. If the mathematics is off-putting, there are general rules that may be followed to avoid the need for calculation, and of course Electronics Workbench provides the ideal way of trying values without the expense of damaged components. **Note:** a small transistor, such as BC108 or BC184, on a 12 V supply, is generally content with a base series resistor of 2.2 kΩ. This will allow the transistor to switch on its maximum current of around 100 mA to 200 mA.

Transistor light switch

Try the circuit shown in Figure 3.2. In this circuit the two potentiometers are connected as variable resistors. This is because one end of the potentiometer is connected in the circuit and the other connection is the wiper.

The lower variable resistor represents an LDR. When the light falling on the LDR increases, its resistance falls. In bright sunlight the LDR will have a resistance of about 200 Ω; in darkness, its resistance will rise beyond 1 MΩ. The voltage at the base of the transistor depends upon the settings of the two variable resistors.

1 Double-click the lamp and set to 1 W. Double-click the lower potentiometer and set to key L (for light).
2 Activate the circuit and press the space-bar to switch on. Try reducing the value of the LDR by pressing key L. At some point

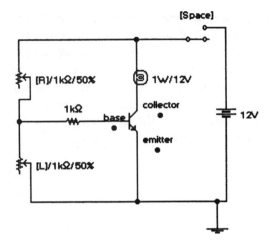

Figure 3.2

the lamp should light. A voltmeter connected between the base of the transistor and 0 V will provide a useful guide as to when the transistor should turn on.

3 Set the LDR to 15 %. Find out what variable resistor setting is required (by pressing R, or SHIFT R) to make the lamp switch on again. **Note**: if the value (%) of the variable resistors cannot be decreased, check that the CAPS LOCK button on the computer is not set.

Explanation

When the LDR is set to a low resistance (say 5 % of 1 kΩ) and the variable resistor is set to 50 %, the voltage at the base of the transistor is just below its turn-on voltage. When the LDR resistance is increased (in real life it could be shaded from light) the voltage at the transistor base rises above the turn-on point and the lamp switches on.

You will have found that the LDR needs to be set to a low value compared with the variable resistor, in order to turn off the lamp. This is because the turn-off voltage at the base of the transistor is very low (less than 0.5 V) compared with the supply of 12 V. It will be easier to adjust the circuit if the maximum value of the variable resistor is changed to 4.7 kΩ. Double-click the variable resistor and change its resistance (R) to 4.7 kΩ. The bulb can now be switched on and off when both the variable resistor and the LDR are around 50 %.

Practical circuit

In practice, the value of the LDR, when shaded, may rise well beyond 1 kΩ and it would be wise to use a higher value variable resistor. As a guide, try 47 kΩ or 100 kΩ.

LDRs are rather slow-acting devices. In other words, when they are shaded their resistance takes an appreciable time to rise (up to several seconds). This is of no consequence in automatic lights or curtain winders, but if you wish to count how many times per second a beam of light is broken by the blades of a fan, the LDR response time will be a problem.

An alternative to the LDR is the photodiode, or phototransistor, which offer much higher operating speeds. They are used in a similar way to the LDR, but ensure that the cathode of the photodiode is connected to the more positive side of the circuit. The collector of the phototransistor must be at the more positive side, its emitter connected to the more negative side and its base left unconnected.

Reversing the action

At present, the lamp lights when the LDR is shaded. If the opposite result is required, the LDR and variable resistor can be interchanged. Note that the maximum resistance of the variable resistor may need to be less than before, depending upon the application.

Using a relay

If the circuit is required to switch a mains lamp on or off, a relay may be employed. The relay coil is connected in place of the bulb shown in Figure 3.2. and the relay contacts can be used like ordinary switch contacts. A relay generates back emf, which can destroy a transistor. Hence, a diode should be connected across the relay coil, with the diode cathode connected to the positive side of the coil.

Capacitor delay circuit

A capacitor is a component that can be charged with electricity. It then stores the charge. It may sound like a re-chargeable battery but it works very differently; a capacitor should not conduct any appreciable current between its two ends.

The amount of charge that can be stored in a capacitor depends upon its value, measured in Farads (F). A Farad is a large unit, and in practice capacitor value is measured in µF (microfarads), nF (nanofarads) or pF (picofarads).

$$1\,\mu F = 10^{-6}\,F$$
$$1\,nF = 10^{-9}\,F$$
$$1\,pF = 10^{-12}\,F$$

Electronics Workbench allows the value of a capacitor to be set by double-clicking. In practice, capacitors above about 1 µF in value are electrolytic types. Electrolytic capacitors are able to squeeze a large value into a small package, but they are polarised – this means that they must be connected the correct way round with respect to positive and negative. Electrolytic capacitors connected the wrong way round may explode after a time!

Real capacitors also have a maximum working voltage. This should not be exceeded. The value of a capacitor may be quite inaccurate; in other words, if the value of a capacitor is supposed to be, say 1000 µF, in reality it may be at least 20 % higher or lower. The value may also change over time, and with temperature. In general, the more you pay, the better the capacitor. Fortunately, the exact value rarely matters in a real circuit.

The capacitor available in Workbench is assumed to be perfect, with no leakage of current between its two ends, an infinite working voltage and non-polarized.

Set up the circuit shown in Figure 3.3. Check carefully that the values of the components have been correctly set up by double-clicking as necessary.

The oscilloscope provides a very useful indication of the action of the capacitor. Drag it from the top of the screen and connect as shown. When the oscilloscope icon is double-clicked it will open. Note that, providing the circuit is earthed, the earth terminal on the oscilloscope need not be connected. However, in real life the earth of the oscilloscope must be connected to 0 V in the circuit.

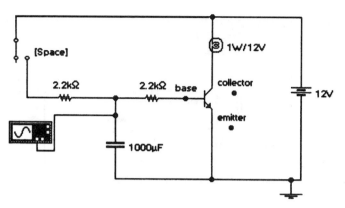

Figure 3.3

Set the timebase of the oscilloscope to 0.5 s/div. Activate the circuit. In all time-delay circuits, you can be misled by delay caused by the software. Keep an eye on the time display at the right of the screen. Any pause on the time display means that the *simulation* has paused – it is not a real delay.

You should notice that, after switching on by pressing the space-bar, there is a genuine pause before the lamp lights. Press the space-bar again: there is a similar pause before the lamp switches off. Observing the oscilloscope trace reveals how the capacitor charges and discharges as the space-bar is toggled.

Further experiments

1 Try changing the value of the capacitor and note the difference in timings and the shape of the trace on the oscilloscope.
2 Replace the lamp with a buzzer. You now have an audible timer. After the timer has sounded, when the space-bar is pressed to reset the circuit there is a delay before the sound stops. Try to arrange the circuit so that the sound stops immediately the circuit is reset.

Hold-on timer

The previous circuit caused a delay before the lamp lit. This example causes the lamp to light immediately, but it then switches off after a time without any further action by the user.

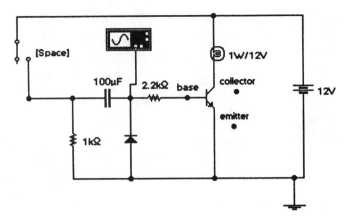

Figure 3.4

1 Set up the circuit shown in Figure 3.4.
2 Check the values of the components carefully, and set the oscilloscope timebase to 0.10 s/div.
3 Activate and press the space-bar. Notice how the lamp lights immediately the space bar is pressed (allowing for delay caused by the software – watch the time counter. Even though the switch remains on, the lamp will switch off after a time.

The action is clearly illustrated by the oscilloscope. The voltage on the transistor side of the capacitor rises instantly, but then decays as the charge leaks through the transistor.

Explanation

It has already been stated that capacitors block the flow of electricity. Yet, in this case, current appears to be flowing from left to right through the capacitor. The explanation is that, although capacitors do block the flow of DC, any change of charge on one side of a capacitor will cause a similar change on the other side. So, if the voltage on the left-hand side rises, there will be a similar rise on the right side.

In a similar way, a charged balloon held near (but not touching) your head will cause your hair to become charged. No current needs to flow between the balloon and your hair, but there is a considerable change of voltage.

Returning to Figure 3.4, the rising voltage on the right-hand side of the capacitor causes a flow of current into the transistor, turning it on. The flow of current causes the voltage to collapse (as shown by the oscilloscope) and, as the voltage falls below about 0.5 V, the transistor switches off.

The 1 kΩ resistor ensures that the voltage on the left-hand side of the capacitor is allowed to fall to zero when the switch is turned off. Try disconnecting one side of this resistor and check the action of the circuit carefully. The first time the space-bar is pressed the time period will be correct. But when the space-bar is toggled again, the time period will be less than expected. Observe the oscilloscope trace for the complete picture.

Reconnect the 1 kΩ resistor and disconnect one side of the diode. Reactivate the circuit and check the oscilloscope. At first everything appears normal, but when the space-bar switch is turned off, the oscilloscope trace moves well below 0 V. This can cause many problems in a circuit! If the voltage at the right-hand side is allowed to fall to near zero before the space-bar switch is turned off, the effect will be even more pronounced. The reason is, again, that if the voltage on one side of a capacitor changes, this change will be copied at the other side. So if the left-hand side is at +12 V, and the right-hand side is at 0 V, when the left-hand side is forced to 0 V, the right-hand side will fall to −12 V.

The cunning use of capacitors can produce voltages in circuits well outside the supply voltage. This is often useful. However, sometimes voltages are achieved that are definitely not wanted, and in Figure 3.4 we do not want the voltage at the transistor base to move far below 0 V. Hence the diode. When the voltage at its cathode end falls below zero, current flows from 0 V (which is now more positive) through the diode, so maintaining the voltage at the cathode end at around 0 V. In practice, this voltage will be about −0.7 V, as there will be a small voltage drop across the diode.

Investigating the transistor

There are two categories of bipolar transistors: 'npn' and its mirror image 'pnp'. Virtually every type of npn transistor has its pnp equivalent. For example, npn type BC108 has a complementary pnp type BC178; type BC184L is the npn complement of BC214L. Starting with npn types, look at the circuit shown in Figure 3.5. The

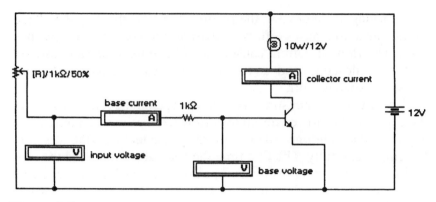

Figure 3.5

potentiometer is used to change the input voltage as shown on the first voltmeter. The second voltmeter shows the base voltage. The two ammeters show the relationship between the base current and the collector current. The bulb is acting as the 'load'. Whether it lights is not important; it is the readings on the meters that are of interest.

Base voltage

Try the circuit in Figure 3.5. One voltmeter shows the voltage between the base and emitter – this will be called 'base voltage'. When the transistor is turned on, this voltage is likely to be around 0.8 V assuming the 'ideal' transistor in Workbench is employed. (The figure of 0.7 V is generally accepted in electronics and physics examinations.)

1 Step the variable resistor to 90 % by pressing SHIFT R.
2 Now double-click the variable resistor and change the 'increment' setting to 1 %. This allows the variable resistor to be changed in 1 % steps.
3 Continue increasing the variable resistor to around 98 %. Notice how the two voltmeter readings are virtually the same. In other words, so little current is flowing into the transistor base that there is virtually no voltage difference across the 1 kΩ resistor. (Remember: Ohm's law states that the voltage difference across a resistor = current × resistance, so if the current is zero, the voltage difference will be zero.)

4 Now step down the variable resistor so that the input voltage reading rises. The base voltage will rise as well, until it reaches around 0.8 V (800 mV). As the input voltage continues to rise, the base voltage will remain at about 0.8 V. More and more current will flow through the 1 kΩ resistor, causing an ever increasing voltage difference. The transistor will ensure that the correct amount of current flows to fix the base voltage.
5 Try increasing the base resistor to, say, 10 kΩ. The transistor will still fix the base voltage to the correct level.
6 Try reducing the base resistor to, say, 10 Ω. The base voltage will still be less than 1 V, even though a massive current is flowing. In real life the transistor is likely to burn out!

It appears that the value of the base resistor is unimportant, but, so far, we have not looked at the current flowing. The need for care in selecting the base resistor value will become apparent when we examine the ammeters.

Transistor current gain

The main function of a transistor only becomes apparent when the ammeter readings are examined.

1 Set the potentiometer to 50 %. Note the base current flowing; note the collector current.
2 Now divide the collector current by the base current. This ratio is known as the *gain* of the transistor. In other words:

transistor gain = collector current/base current

For example, if the collector current is 200 mA and the base current 4 mA, then:

gain = 200/4 = 50

3 Reduce the potentiometer to 40 %, and note the new values on the ammeters. Calculate the gain.

Conclusion

The ratio of collector to base current is roughly the same and the transistor *amplifies* the base current. Hence, a small current delivered to the base can be used to control a much larger current through the transistor.

What happens to the current delivered to the base? Insert an ammeter into the lead joining the emitter to 0 V. What do you notice about the emitter current compared with the base and collector currents?

Saturating the transistor

So far we have ignored the effect of the bulb in the circuit and, with ratings of 10 W 12 V, it passed sufficient current not to upset our earlier calculations.

1 Try reducing the power rating of the bulb to from 10 W to 1 W.
2 Set the potentiometer to 30 %.
3 Calculate the transistor gain.

This time, the gain will be much lower than before – the transistor is *saturated*. In other words, the transistor is turned fully on, and the collector current is being limited only by the bulb. Reducing the setting of the potentiometer will not affect the collector current. However, if the potentiometer is raised to 80 %, thus reducing the input voltage to a very low level, the collector current will start to decrease – the transistor is moving below its saturation point.

Calculating the value of the base resistor

Assuming that the transistor is to be used as a switch, we need to apply sufficient base current to saturate the transistor. The maximum collector current that the transistor can tolerate can be looked up in the supplier's catalogue. The gain of the transistor can be established in the same way, except that transistor gains can vary enormously, even in transistors of the same type. Hence, your catalogue will print the *lowest* gain expected. Circuit designs based on this figure will also work if the transistor gain is higher. Transistors also have a maximum working voltage, but this is generally well above 12 V and so need not concern us.

Assuming that the transistor has a gain of 100 and a maximum collector current rating of 200 mA, as:

gain = collector current/base current

then:

base current = collector current/gain

so:

base current = 200 mA/100 = 2 mA

In other words, the base current must be 2 mA for saturation.

Assuming that the emitter is connected to 0 V, the base voltage will be around 0.7 V. (It was actually 0.8 V in the Workbench circuits, but 0.7 V is a figure agreed by virtually all examination boards.) So, the voltage difference across our base resistor will be the input voltage less 0.7 V. If the input rises to 12 V, the voltage difference will be 11.3 V. Using Ohm's Law, we can now calculate the base resistor value.

$$R = V/I$$

So:

$$R = 11.3 \text{ V}/2 \text{ mA} = 5.65 \text{ k}\Omega$$

Note: If the value of the base current is in mA (i.e. not changed to 0.002 A), then the answer will be in kΩ.

As a resistor value of 5.65 kΩ does not exist, a lower value that will still provide full saturation must be chosen. This could be 5.6 kΩ, but, in practice, any resistor value down to about 1 kΩ will be satisfactory, and a lower value will ensure proper saturation even if the supply voltage is reduced slightly. Of course, the value must not be too low, or the transistor will be burnt out by excessive base current.

Pnp transistors

Try the circuit shown in Figure 3.6. Notice that, to turn on the transistor, the input must be connected to 0 V, rather than the

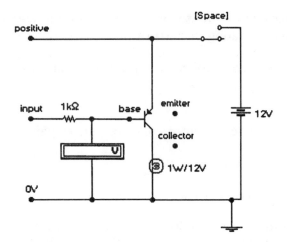

Figure 3.6

positive supply rail. The voltmeter will indicate a reading of at least
0.7 V *less* than the positive supply.

The circuit in Figure 3.7 shows the pnp equivalent of Figure 3.5.
Try the circuit. Note that the base current is *negative* (as indicated
by a minus sign). This shows that the current is flowing *out* of the
transistor base.

All the facts and figures for the npn transistor apply equally to
the pnp type. Avoid confusion by remembering that the arrow
always indicates the *emitter*, and its direction shows the conventional flow of current (conventional meaning from positive to
negative).

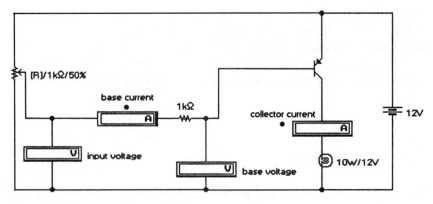

Figure 3.7

The transistor circuits described so far are known as 'common emitter' circuits. This is because the emitter is common to the input and output sides of the circuit. There are other ways in which transistors may be connected, and the next circuit shows a configuration known as an 'emitter follower'.

Voltage controller

A voltage controller can be employed as a lamp dimmer, motor speed controller, etc. An attempt to control voltage was seen in Chapter 1. However there were problems: a small potentiometer can only be employed if the current required is very small – say 10 mA or less. A potentiometer connected to make a variable resistor will only control voltage if a known current is flowing and, again, a small potentiometer will overheat if too much current flows.

What is required is a device that will amplify the tiny current available from a standard potentiometer. A transistor connected as an 'emitter follower' provides the answer.

Try the circuit shown in Figure 3.8. **Note:** the power of the bulb has been reduced to 0.1 W. The potentiometer has been rotated. This means that the output from its wiper (shown by an arrow) *rises* when the percentage shown is *reduced*.

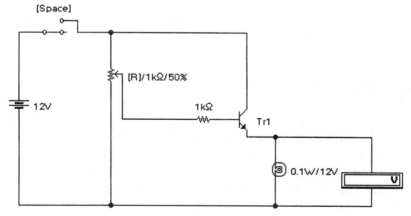

Figure 3.8

Activate the circuit, and switch on by pressing the space bar. The brightness of the bulb is indicated by the reading on the voltmeter. As the variable resistor is moved towards 0 %, the voltage at the transistor base will rise.

We saw earlier that there is a voltage difference between base and emitter of about 0.7 V to 0.9 V. Therefore, if we fix the base at a particular voltage, the emitter will 'follow' this voltage to within about 0.8 V – hence the name emitter follower.

The advantage is that the current available from the emitter will be many times the current required at the base. The default transistor in Workbench provides a current gain of 49 – in other words, if 49 mA is supplied by the emitter, the base will require only 1 mA. This current can be supplied easily by a small potentiometer.

In summary, the potentiometer sets the voltage and provides a small current; the transistor amplifies this current to a useful level and delivers it from the emitter at a voltage of about 0.8 V less than the output from the potentiometer. Hence, the circuit can supply any voltage from zero to about 0.8 V less than the supply voltage.

The fixed 1 kΩ resistor is required to limit the current flowing into the base to a safe level should the potentiometer output be set to maximum. As the current flowing is small, the fixed resistor does not greatly affect the voltage applied to the transistor base.

Practical points

A small transistor, such as BC108, will provide a gain of at least 100. A BC184 will achieve a gain of at least 200. Both these transistors can be damaged if too much current is supplied from the emitter; avoid currents in excess of 100 mA.

Greater power

Most voltage control circuits are required to provide more than 100 mA. However, it would be unwise to use a larger power transistor since the gain of such a transistor might be as low as 20. This would require a much larger current at the base – sufficient to damage a small potentiometer.

The solution is provided by a darlington pair: a pair of transistors consisting of a small high-gain type combined with a larger high-

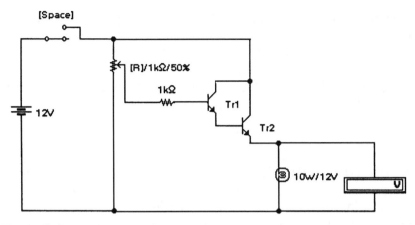

Figure 3.9

power type. This results in high gain; high power – the best of both worlds!

Figure 3.9 shows the arrangement. Transistor Tr1 is a small npn high-gain type, such as BC108 or BC184. Tr2 is an npn power transistor, such as TIP41A. The total gain provided by the darlington pair is found by multiplying the gains of the individual transistors. So if Tr1 has a gain of 100 and Tr2 a gain of 30, the total gain is 3000.

To check the gain figure, return to Figure 3.5 and add a second transistor to make the darlington pair arrangement shown in Figure 3.9. Increase the power of the lamp to 100 W. As the gain of the default transistor is 49, the total gain of the pair should be about 2401 (49 × 49).

If the voltage between the base of Tr1 and the emitter of Tr2 is measured, there will be a reading of about twice the base/emitter voltage for one transistor. This is generally of no consequence, but in the voltage control circuit it means that the maximum output voltage is around 1.5 V *less* than the supply voltage. The maximum current available is the current that the power transistor can handle – this is about 6 A with a TIP41A.

It is possible to purchase a single darlington, that is, a transistor that looks like an ordinary power transistor, but which also includes a low-power high-gain type. For example, type TIP121 is a darlington pair with a gain of at least 1000 and maximum current handling of 5A.

Mains voltages

The voltage controller is *not* suitable for mains use, unless the mains voltage is isolated and reduced by means of a transformer, as described in Chapter 4. Mains dimmer circuits will be described in Chapter 14.

Motor reversing circuit

We saw in Chapter 2 how switches can be used to control the direction of a motor. A relay with two pairs of change-over contacts (i.e. DPDT contacts) can be used in a similar way. However, the transistor circuit shown in Figure 3.10 provides a useful alternative. There are no contacts to burn out and the circuit operates silently.

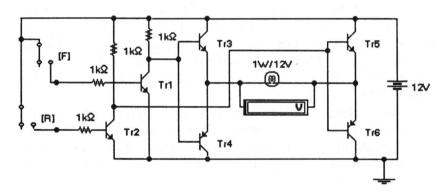

Figure 3.10

Transistors Tr1 & Tr2 are npn types, and in a practical circuit could be low-power high-gain types, such as BC108 or BC184. These transistors enable the circuit to respond to very small amounts of current. The other transistors are high-power types, and note that Tr4 and Tr6 are pnp transistors. In a practical circuit, Tr3 & Tr5 could be TIP41A and Tr4 and Tr6 could be TIP42A. In the Workbench circuit, use the default transistors, as this helps the software to operate more quickly.

As usual, the bulb indicates that current is flowing, and the voltmeter indicates the direction. Do not be surprised if nothing happens when a switch position is changed: the circuit is sufficiently complex to slow down all but the fastest computers and a frozen time display indicates that the computer is still calculating the result. Be patient!

Further work

Try substituting LEDs for the bulb. Connect them in 'inverse parallel' as shown in Figure 2.4; do not forget a series resistor to limit the current to a safe level. Note that the use of LEDs will slow down the simulation.

Practical points

Motors are electrically noisy devices: they generate voltage spikes (due to back emf) and these spikes can destroy transistors and upset the functioning of any controlling circuit. Figure 3.11 shows how four diodes and a capacitor are used to reduce the problem. These components were not included in Figure 3.10 as they tend to slow down the simulation. The diodes are type 1N4001, and the capacitor is a 100 nF (0.1 µF) ceramic-disc type.

The need for good decoupling (explained in Chapter 15) is especially important in this type of circuit.

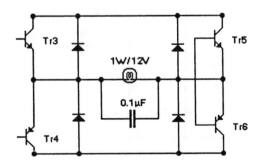

Figure 3.11

Applications

This circuit is ideal for controlling a motor from a logic control circuit. For example, an automatic curtain winder might include a light sensor, an infra red remote control sensor, and push buttons, all linked to a control unit based on logic gates. Such a circuit is shown in Chapter 5.

Transistor amplifier

As a transistor is a current amplifier it would seem well suited to being made into an audio amplifier. However, if we wish to amplify voltage we must ensure that the base of the transistor is moved between about 0.5 V to 0.9 V higher than its emitter. If the base voltage moves outside of this range, *distortion* will occur, as the transistor will either be switched off, or saturated.

Modern amplifiers almost invariably employ integrated circuits (ICs) in their design. Although ICs consist mainly of transistors (in integrated form) they are more stable and more reliable, and are less prone to 'thermal drift' – varying performance with changes of temperature. We will look at IC amplifiers in Chapter 5.

A single transistor amplifier is shown in Figure 3.12. It is known as a common emitter amplifier and can amplify AC (or audio)

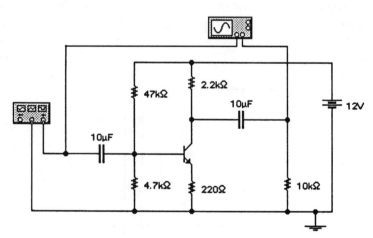

Figure 3.12

signals. This signal is obtained from a signal generator, located at the top of the screen.

1 Pull down the signal generator, double-click and change its frequency to 1 kHz. and its amplitude to 50 mV.
2 Pull down and double-click the oscilloscope. Change its timebase to 0.20 ms/div and change the inputs of both channel A and channel B to 200 mV/Div. Notice that the oscilloscope is connected directly to the signal generator to show the input signal, and connected to the output of the amplifier to show the amplified signal.
3 Double-click on the wire leading to the oscilloscope and change its colour. The oscilloscope trace will change colour to show which is which.

When the circuit is activated the oscilloscope should reveal an amplified signal that is out of phase with the input. In other words, when the input signal rises the output signal falls. This phase difference is of no consequence in audio amplifiers.

Practical circuit

It must be stressed that a simple transistor circuit has severe limitations and much better amplifiers are possible with integrated circuits. But this circuit is useful for experimentation.

The audio signal might be the output from a microphone. The 10 kΩ 'load' resistor represents the device through which the signal is required – the signal would normally be delivered to a power amplifier circuit, in which case the load could be raised to 1 MΩ. If headphones are used to test the output, the signal will be reduced according to the impedance of the headphones. (Impedance can be likened to resistance, for now.) Try reducing the 'load' to 100 Ω. The output signal will be much less.

A full explanation of this circuit (including the choice of resistor values) is beyond the scope of this book, but the principle is that the DC voltage at the transistor base is fixed to allow the collector voltage to 'sit' at about half the supply voltage. When an AC signal is applied at the input, the transistor base voltage 'wobbles' up and down, causing its collector voltage to wobble *down and up*. Note that the emitter is connected via a resistor to 0 V, so that it can sit at the correct voltage. In fact, the emitter voltage also wobbles up and down with the base. Try connecting a 1 µF capacitor between the emitter and 0 V: the output signal is larger. However, if the capacitor is made too large in value, then the output may become distorted.

Push-pull power amplifier

It should be noted that, like the previous transistor amplifier, this power output stage suffers from problems, which are most easily cured by using a specialized integrated circuit. However, the module shown in Figure 3.13 does illustrate how npn and pnp transistors can be made to work together.

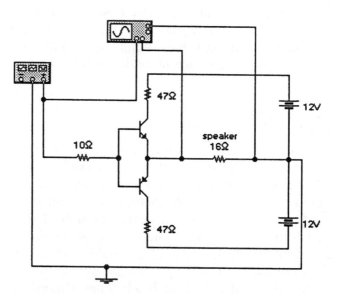

Figure 3.13

Dual rail supply

The power supply is split as shown with the 'ground' connection joined at the point where the two batteries connect. This type of supply is commonly used in amplifiers, particularly if they are driven from a mains supply via a transformer. The supply delivers a positive 12 V supply, a 0 V rail, and a negative 12 V supply. This enables the circuit to drive a loudspeaker in the most efficient way; that is, the loudspeaker cone is pushed equally forwards and backwards about its mid-point. As before, the signal is derived from a signal generator.

1 Set the frequency to about 1 kHz and the amplitude to 2 V.
2 Set the oscilloscope timebase to 0.20 ms/div and channel A and channel B to 1 V/div.
3 Set the oscilloscope input leads to different colours so that the traces can be identified.
4 Activate the circuit. Notice how the output signal is lower in amplitude than the input; the amplifier does not appear to amplify! However, the current available for the speaker is much greater than the input current, and this is the main purpose of a power amp designed for driving low-impedance speakers.

Crossover distortion

The output signal suffers from a form of distortion known as 'crossover distortion'. This is due to the fact that the emitter voltage of an npn transistor is about 0.7 V lower than the base, and the emitter voltage of a pnp transistor is about 0.7 V above the base. So, when the base voltage is between about −0.7 V and +0.7 V, both transistors are switched off.

Practical amplifier

If this were made into a real circuit, the transistors should be power types, such as TIP41A (npn) and TIP42A (pnp). The 47 Ω resistors will be passing a very large current and should be rated at 5 W to avoid overheating.

Questions

Full written answers complete with Workbench circuits, are available on the accompanying disk. See p. 224 for details.

1 Using the circuit in Figure 3.2 as a starting point, design a transistor circuit that sounds a buzzer whenever an LDR is

shaded from light. (The LDR can be represented by a variable resistor labelled L.)

2 Now remove the buzzer in Q. 1, and replace it with a relay coil. Select another battery, but change its voltage to 230 V. Select a lamp and change its voltage to 230 V to match the new battery. Connect up the new battery and bulb to the relay contacts in such a way that the bulb lights whenever the LDR is shaded. Add a diode to protect the transistor against back emf from the relay.

3 Using the circuit in Figure 3.3 as a guide, design a timer that sounds a buzzer after a delay of about 3 seconds.

4 Draw a circuit with a lamp (10 W, 12 V) connected between a transistor collector and positive. Use a variable resistor to control the current flowing into the base of the transistor. Employ a 12 V battery. If the transistor has a gain of 50, calculate the current that must flow into the transistor base to allow a current of 600 mA to flow through the bulb. Check your answer by including ammeters to measure the collector and base currents.

5 Design a voltage controller circuit for a model train set. Use a darlington pair, and assume that your train requires a supply of between 0 and 12 V at up to 1 A. Your voltage controller is powered from a mains power unit with a 15 V, 1 A DC output. Use a 12 W, 12 V bulb to represent the train.

Chapter 4

Power supplies

All projects need a power supply. This might be just a battery, or a battery circuit with a voltage regulator. Voltage regulators are available as integrated circuits, and Workbench allows us to create our own IC and store it in the customized parts window.

We will begin with the method by which AC mains supplies are converted into a DC supply suitable for driving all the projects in this book. It must be stressed that no person should tamper with a mains supply unless suitably qualified; cheap AC adapters are available that provide a low voltage output.

Transformer

We will begin with the mains transformer. Figure 4.1 shows a suitable testing circuit. The default mains supply is 120 V, 60 Hz AC. This can be changed (by double-clicking on the supply) to 230 V, 50 Hz AC – the European mains supply. Just to add to the confusion, the UK mains supply *used* to be rated at 240 V, but the new rating conforms with the European standard.

Try the circuit shown in Figure 4.1. Double-click on the voltmeters and change them from DC to AC. Double-click on the transformer and select a 'Power 10 to 1' ratio. Note that the transformer performs two jobs: it isolates the mains supply, so that the risk of electrical shock is much reduced, and it lowers the voltage to a safe 12 V.

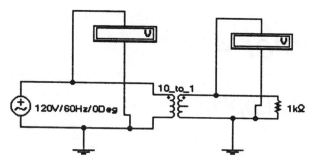

Figure 4.1

The ratio of the turns on the transformer determines the output voltage. A ratio of 10:1 will reduce the input voltage by a factor of 10. However, we have not lost energy, as the current required by the mains coil will be 10 times less than the current supplied by the low voltage coil.

Try changing the mains supply to the European standard of 230 V, 50 Hz. Also, change the transformer to a ratio of 25:1. The resulting output voltage should be about 9 V. This is quite a useful figure, as we will see later.

Centre tap

Try connecting the top of the right hand voltmeter to the centre tapping of the transformer: the voltage should be half the previous value. The usefulness of the centre tap will become clear when we look at complete mains power supplies.

Peak voltage

The voltmeters – when set to AC – display the readings we might expect. But if the voltmeters are replaced with the oscilloscope, it is clear that the situation is far from simple!

1 Try connecting the oscilloscope as shown in Figure 4.2.
2 Set its timebase to 5.00 ms/div and channel A and channel B to 50 V/div.

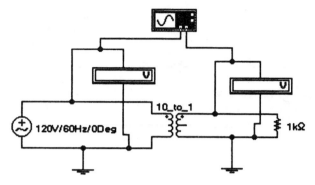

Figure 4.2

3 Check the amplitude of the input supply against the scale on the oscilloscope screen.

The input voltage of, say, 120 V is actually peaking at 170 V and the 12 V supply is peaking at 17 V. So the voltmeter readings were a sort of 'average'. A true average for an AC supply would be zero, as AC moves equally between positive and negative. The average shown by the voltmeters is 'RMS average', where RMS stands for 'root mean square'. The RMS value is obtained by the formula:

$$\text{RMS} = \text{peak value}/\sqrt{2}$$

So that:

$$\text{peak value} = \text{RMS} \times \sqrt{2}$$

(note that $\sqrt{2} = 1.4$ approx.)

Hence, the UK mains supply of 230 V actually peaks at 325 V. The importance of noting the peak value of the output voltage will become clear when we look at a complete power supply circuit.

Diode rectifier (half wave)

Rectifiers are circuits designed to change an AC supply into DC. Our mains supply is AC and it is very easy to convert AC to a higher or lower voltage using a transformer, as shown above.

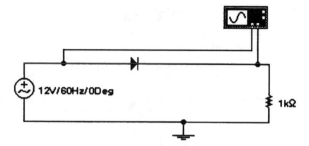

Figure 4.3

However, nearly all electronic circuits require DC and diodes are used to change AC to DC.

1 Try the circuit shown in Figure 4.3. Note that the AC voltage source has been set to 12 V.
2 Set the timebase of the oscilloscope to 5.00 ms/div.
3 Double-click the wires leading to the oscilloscope and set them to different colours.
4 Activate the circuit.

Note that one trace on the oscilloscope shows the AC voltage before the diode, and the other trace shows the output from the diode. The output voltage is DC, as it only moves *above* the centre line of the oscilloscope, but it is not the *same* as the steady DC supplied by a battery. The type of DC supplied by a single-diode rectifier is known as 'unsmoothed DC'. Note the large gaps in the output during the times when the AC supply is reversed.

The resistor represents the load. This could be any circuit that requires a DC supply. Without a load, i.e. if no current flows through the diode, odd results will be observed on the oscilloscope, both in Workbench and in real life.

Further work

If unsmoothed DC is supplied to an amplifier or a similar circuit that requires a smooth DC supply, severe distortion or other malfunction will occur. Try to invent a simple way (i.e. using one extra component) to smooth the DC supply in Figure 4.3. The answer is provided later in this chapter.

Bridge rectifier (full wave)

A single diode resulted in large gaps in the DC supply. Four diodes arranged as shown in Figure 4.4 will allow these gaps to be filled. The bridge arrangement is available ready-assembled in the parts bin. In real life, bridge rectifiers are available ready-made, though it may be a little cheaper to use four diodes.

Try the circuit shown in Figure 4.4. Set the oscilloscope time base to 5 ms/div. Observe the shape of the trace – it is above the 0 V (centre) line, i.e. it represents DC. Where the circuit in Figure 4.3 caused a gap between the waves, the gaps are now filled in. The full AC wave is now used, but it has been changed to DC, although it is still unsmoothed DC. Try to find which component from the parts bin will smooth the wave so that only a small 'ripple' remains.

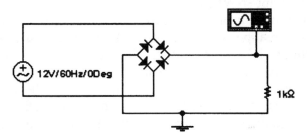

Figure 4.4

Note: it is not possible to show the AC input and the DC output on the oscilloscope at the same time, as the oscilloscope 'earth' cannot be connected to *both* the AC input and the DC output. The same applies in real life.

Full wave power supply

Note: This section shows the construction of a mains power supply. It is intended for information only, as mains supplies are dangerous and no person without the appropriate qualification should tamper with the mains. Inexpensive mains supplies are available ready made, and these are recommended.

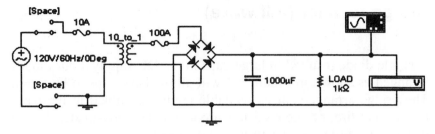

Figure 4.5

We are now ready to employ all the components to create a mains-driven power supply that will convert our mains supply from high voltage to a safe low voltage and also change it from AC to DC.

Try the circuit shown in Figure 4.5. Note the use of fuses and a double pole switch: these are safety devices and must always be included in a real power supply. A neon indicator on the mains side of the circuit is also essential. Note that the fuse ratings in Workbench must allow for the surge of current into the capacitor when the circuit is first switched on. In real life, the fuse ratings should be close to the normal maximum current expected. Real fuses will pass many times their maximum current rating for a very short time. The switches and fuses may slow down the simulation in Workbench and can be omitted if necessary.

Set the timebase of the oscilloscope to 2.00 ms/div.

Output voltage

The DC output may be higher than expected. Remembering that the transformer in Figure 4.5 is supplying an AC output of 12 V, we appear to be gaining voltage. The output of 12 V from our transformer is an 'RMS value. Hence, the peak value is 1.4 times higher. The trace on the oscilloscope shows that the DC output rises to the peak value, assuming that a suitable capacitor is fitted in parallel with the load as shown.

We would thus expect a DC output of 1.4 × 12 V, i.e. 16.8 V, but this is not the case! Each diode causes a loss of about 0.7 V. As, at any one moment, two diodes are forward biased and therefore conducting, there will be a loss of about 1.4 V. So, our final output voltage will be about 16.8 less 1.4, i.e. 15.4 V. This is still higher than the original supply of 12 V AC and must be allowed for when designing a power supply.

Current

The rise in voltage is compensated for by a loss of current – we cannot create energy from nowhere! Hence, if the voltage rises by 1.4 times, the current available will be correspondingly smaller, i.e. the current supplied by the transformer will have to be 1.4 times greater than the current used by the load. For example, if the load requires a current of 1 A, you must ensure that the transformer can supply $1 \times 1.4 = 1.4$ A.

Experiments

Our load of 1 kΩ allows a current of around 10 mA to flow. This is very small and, in practice, a power supply would normally be expected to supply much more current. Try reducing the value of the load resistor to 10 Ω; this will conduct a current of about 1 A. Observe the oscilloscope trace: a pronounced ripple has appeared on the DC supply. This might cause severe problems to some circuits, e.g. an amplifier, radio or tape recorder driven from such a supply, will produce an unpleasant hum through the speakers.

Try reducing the ripple by increasing the value of the capacitor. This should work, but capacitors greater than about 10 000 μF are large and expensive and a large amplifier will require a current of 10 A or more. Try reducing the load resistor to 1 Ω, then check how large the capacitor value must be for an acceptable level of ripple.

Manufacturers of powerful amplifiers have been tackling this problem for many years, and there is no simple solution. In fact, most of the weight and size of a domestic amplifier unit is accounted for by its power supply. Later we will see how the problem of ripple can be entirely eliminated, but this only works for small currents.

Alternative full wave power supply

The term 'full wave' is used to describe a power supply that uses both the positive and the negative of the AC supply to create the DC output. The power supply above achieved this by employing a bridge rectifier.

It is possible to achieve the same result by using the centre tapping of the transformer. Not all transformers have a centre tap, but if they do, they are often described as, e.g. 6–6. Sometimes, transformers have two secondary coils, which may be connected in series to create a centre tap. Such transformers may be described as, e.g. 0–6, 0–6, indicating two separate 6 V coils.

Figure 4.6 shows how only two diodes are needed to create the full wave supply. Try the circuit, and prove that it really is a full wave supply by removing the smoothing capacitor (don't leave a short-circuit when the capacitor is removed).

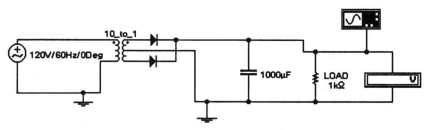

Figure 4.6

Output voltage

The output will be based upon the voltage across each half of the secondary coil. In this case, the transformer provides 6 V AC (that is, 6 V RMS). So the peak output voltage will be 6 × 1.4 = 8.4 V. At any one time, a single diode will be forward biased and will remove about 0.7 V, hence, the expected DC output, assuming that a smoothing capacitor is employed, will be about 7.7 V.

We appear to have lost voltage, as the transformer was capable of supplying 12 V AC. However, the DC current available will be correspondingly higher.

Variable voltage supply

Try the circuit shown in Figure 4.7. You will see that it is simply a combination of the power supply circuit shown earlier in this chapter combined with the voltage controller from Chapter 3.

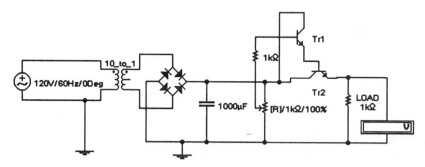

Figure 4.7

The load resistor represents the circuit or device being powered by the voltage controller. It may be a lamp, motor, etc. There is no need to fit a resistor to represent the load, except for testing purposes. Note that if 'inductive loads' are used – that is, devices such as motors, which include a coil of wire – then it is wise to fit a diode reverse biased across the output. In other words, the cathode end of the diode must be connected to the positive output side. The diode will not normally conduct any current, but if the motor generates a high voltage spike, the diode will conduct it safely away before it destroys the transistors.

In real life...

The hazards involved in dealing with mains supplies should not be underestimated, and an inexpensive AC adapter with DC output could be purchased ready built. The transistor controller circuit could then be added to provide a range of outputs from zero to a little less than the DC input voltage. Tr1 could be a BC108 or BC184 and Tr2 could be a power transistor TIP41A.

The two transistors, in fact, form a darlington pair, and a single darlington transistor could be purchased, such as a TIP121.

The circuit provides voltage control via a small potentiometer (like the type used in volume controls). The output voltage is not regulated, i.e. it is likely to fluctuate according to the current drawn by the load. This may or may not be a problem, but the next circuit shows how the output can be fully stabilized.

Stabilized supply

As the name suggests, the purpose of this circuit is to provide a stable voltage (often known as a regulated voltage) regardless – within reason – of the current used by the load. We will also see, for the first time, how we can make our own Workbench integrated circuit (or chip).

Ohm's Law describes how voltage, current and resistance are interdependent. If the resistance remains constant and the current changes, then the voltage must change as well. So, if we want the voltage to remain constant even when the current changes, we must find a device which does *not* obey Ohm's Law.

The diode is one such device. When forward biased (i.e. so that current will flow) there is a fairly steady difference across the diode of just under 1 V (if the diode is made with silicon). When reverse biased, no current flows until the voltage exceeds the PIV rating of the diode (see Chapter 2).

Zener diode

Some diodes are manufactured with a low, and accurate, PIV. These are known as zener diodes.

Try the circuit shown in Figure 4.8. Double-click the zener diode and select type 1N4733. The load is represented by a potentiometer (connected to make a variable resistor). In real life the load might be a sensitive logic circuit driving a seven-segment display (the type that lights up different segments according to the required digit). The logic circuit requires say, 5 V, but the supply available is rated at 12 V. How can we reduce a 12 V DC supply to 5 V, bearing in mind that the current requirements will change according to the digit being displayed? A simple series resistor could be employed, but the voltage drop across the resistor will vary according to the current flowing.

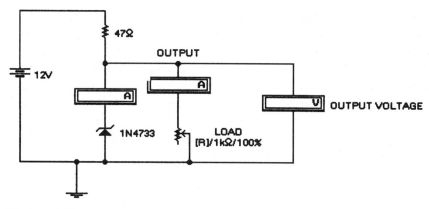

Figure 4.8

The zener diode circuit in Figure 4.8 solves the problem. Try changing the current through the load by changing the variable resistor percentage – move the percentage between 5 % and 100 %. The current through the load (as shown on the ammeter just above it) will change by a large amount. But the voltage across the load (as shown on the voltmeter) will stay at around 5 V. Even if the percentage load is set at 5 % the output will remain at above 5 V. At the other extreme, if one end of the load is disconnected, so that no current flows at all, the output will not rise above about 5.3 V. Hence, the output current can move from zero to just over 100 mA with little change of output voltage – a well-regulated supply! Try changing the battery voltage. Providing it is above about 10 V its actual value will make only a minor difference to the output voltage.

How the circuit works

The zener diode is connected in reverse biased mode. In other words its cathode is connected to the more positive side of the circuit. This means that current will not flow through the diode unless the voltage across it exceeds a certain value, known as the zener breakdown voltage. This special voltage is fixed when the diode is manufactured – you select a particular voltage when you buy the diode.

The zener diode selected in Figure 4.8 has a voltage setting of 5.1 V. You can check this by double-clicking the zener diode then clicking on EDIT. The diode test voltage is 5.1 V at a stated current.

If the current changes, the zener voltage will remain more or less constant.

Try Figure 4.8 again, checking the readings on the two ammeters while changing the percentage of the load resistor. Notice how, if the load resistor is at 5 % (i.e. a low resistance) most of the current flows through the load. If the percentage is increased, more and more current will flow through the zener. If the load is disconnected so that the current falls to zero, all of the current through the 47 Ω resistor will flow through the diode.

Zener limits

It is possible to vary the current through the load in Figure 4.8 to a maximum of about 100 mA. This current is limited by the 47 Ω resistor. If you require a higher maximum current, this resistor can be reduced in value. The zener diode will still conduct sufficient current to maintain a difference of about 5.1 V across its two ends. If the load is now disconnected, all this current must flow through the zener.

Try this by reducing the resistor to say 2 Ω (make sure the battery is returned to 12 V). The poor zener diode will be conducting just over 3 A. In real life it will have been destroyed.

Zener diodes have a certain power rating. For example, if a zener is rated at 1.3 W, the maximum current that can be tolerated with a 5.1 V zener is given by:

$$\text{current} = \text{power/voltage}$$
$$= 1.3/5.1 = 255 \, \text{mA}$$

Calculating the resistor value

In our example, with a 5.1 V zener and a 12 V supply, the voltage across the resistor will be $12 - 5.1 = 6.9 \, \text{V}$. We have already calculated that the maximum permissible current through the zener diode is 255 mA (0.255 A). So, using Ohm's Law:

$$\text{resistance} = \text{voltage/current}$$
$$= 6.9/0.255 = 27 \, \Omega$$

Return to Workbench and try a 27 Ω resistor. Allowing for the imperfections in the zener diode (which are simulated by the software) the total current that the zener must conduct if the load is disconnected will be roughly the expected value.

There is one step remaining before attempting to build the circuit. We often ignore the power handling requirements of the resistors in circuits, because the current flowing is so small that there is little risk of them overheating. But in this case we *must* check the power rating of the resistor. The power that the resistor is required to dissipate depends upon the voltage across it and the current flowing through it:

$$\text{power} = \text{voltage} \times \text{current}$$

So, in our case:

$$\text{power} = 6.9 \times 0.255 = 1.7\,\text{W}$$

Hence, our normal 0.25 W resistor will not do! A larger and more expensive 2 W resistor is required.

Improvements

Our circuit is far from perfect. We cannot buy a zener diode with a large power handling capability, and even if we could, the resistor would also have to be large and expensive. So our output current is severely limited.

The solution is to employ a transistor. Not only will this transform our zener circuit into a very useful voltage regulator, but we can make it into our first Workbench integrated circuit, or chip.

Transistor/zener regulator

Try the circuit shown in Figure 4.9. Note that the value of the resistor in series with the zener diode has been raised to 100 Ω. Hence, the current flowing through the zener is well within its limit. However, the transistor is used to amplify the current available (in this case by about 50 times), so that nearly 1 A can flow from the output.

Check that the load resistor has been changed to 100 Ω. When this is set to 5 % the current flowing is nearly 1 A (1000 mA). The output voltage is still well regulated, but it is about 0.7 V less than with the circuit shown in Figure 4.8. The voltage loss is caused by the difference in voltage between the base and emitter of a

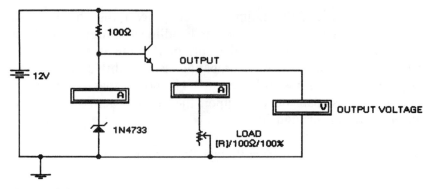

Figure 4.9

transistor. If the base is held at a particular voltage (by the zener in this case), the emitter will be at about 0.7 V less (this difference was discussed in Chapter 3). So a 5.1 V zener will result in an output of about 4.4 V.

Double-click the zener diode and select an alternative that will provide an output of as close to 5 V as possible. The zener diode and transistor are part of a circuit that can be purchased as an integrated circuit (IC). For example, regulator type 7805 is a three-pin device, which looks like a power transistor, and provides an accurate 5 V output from any input between 8 V and 36 V.

Creating your own IC

Workbench allows the creation of 'subcircuits' – a group of components enclosed to resemble an IC. Begin with the circuit shown in Figure 4.9.

1 Remove the ammeter in series with the zener diode (but leave the zener connected to the resistor and transistor).
2 Use the mouse to select the zener diode, series resistor and transistor. Check that no other components have been highlighted by mistake.
3 Click on CIRCUIT at the top of the screen and select SUBCIRCUIT.

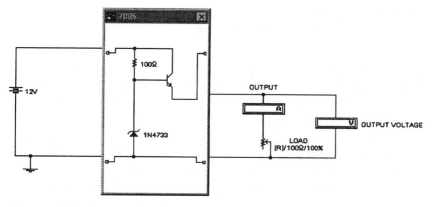

Figure 4.10

4 Type in the name 7805 (or any other of your choice) and click COPY. The box will appear as in Figure 4.10. The small squares at each corner of the circuit are connectors.

5 Move them so that they resemble the pattern shown in Figure 4.11. For example, if you pull one of the lower squares to the bottom centre, then pull the other lower square out of the box – it should vanish.

6 Close the box – the subcircuit will vanish into the custom parts bin. Click on the Custom parts bin (to the left of the resistor icon) and you will see the 7805 symbol. Whenever this particular

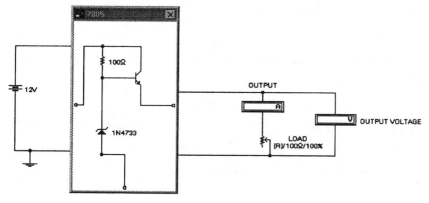

Figure 4.11

circuit is loaded, the 7805 should appear in the Custom parts bin, and you can use as many times as you like.

At any time, the 7805 symbol can be double-clicked to show its contents and it can then be modified as required. Try changing the zener diode type for one that produces an output of 12 V. The new subcircuit should be called 7812.

Note: remember, when experimenting with higher voltage zener diodes, that the supply voltage must be several volts higher than the expected output. Raise the battery voltage if necessary.

In real life the 7805 regulator produces an output of 5 V at up to 1 A. The 7812 produces 12 V at up to 1 A and the 7815 produces 15 V at up to 1 A. All three regulator ICs are known as 'positive' supply types, and they employ npn transistors. A set of three regulators that begin with '79' are also available – these are 'negative' supply types.

Another group of regulator ICs begin with '78L' (and '79L'). These are similar, except that they can supply a maximum of 100 mA. This makes them very valuable when testing a small circuit for the first time, as even the worst mistake can only result in a current of about 100 mA. This is insufficient to cause much damage.

Creating a 9 V supply

A regulated 9 V supply can be very useful, but 9 V versions of the ICs mentioned are not generally available. Figure 4.12 shows how a resistor can be added to raise the voltage of the 5 V regulator. The

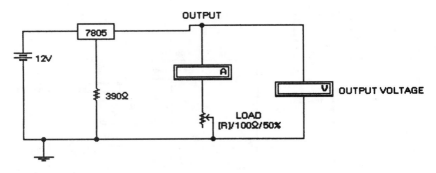

Figure 4.12

value of the resistor determines the increase in voltage. The supply will not be as well regulated, but most equipment will happily tolerate a volt or two on either side of 9 V.

Questions

Full written answers, complete with workbench circuits, are available on the accompanying disk. See p. 224 for details.

1 A transformer with a turns ratio of 20:1 is used to reduce an AC mains supply of 230 V. Find the expected output voltage.

2 If the RMS output from a transformer is 12 V, find the peak output voltage.

3 Show how a zener diode with resistor may be used to supply a fixed 4.7 V output. Use a 12 V battery as the power supply. Select the resistor value so that you can obtain a maximum of 15 mA at the output.

4 Using the figures supplied in Q. 3:

(a) calculate the zener diode power rating required
(b) calculate the resistor power rating needed.

5 Show how a transistor (with a gain of 50) may be employed to boost the available current in Q. 3 to about 1 A. Select a zener diode that provides an output from the transistor as close as possible to 6 V.

6 Design a power supply that can be plugged into the UK mains supply of 230 V, and which produces a smooth DC output of 12 V. Take the following steps:

(a) draw the circuit
(b) calculate the AC output voltage required (i.e. the RMS value) from the transformer that causes 12 V to develop across the smoothing capacitor
(c) calculate the transformer turns ratio required
(d) check your calculations on Workbench.

Note: you should select the POWER IDEAL transformer (by double-clicking on the symbol). Now click on EDIT and change the turns ratio to the value you have calculated.

7 Add a resistor, zener diode and transistor to your circuit in Q. 6 that will regulate the output and provide a supply of about 5 V at up to about 100 mA. State the value of the resistor required.

Chapter 5

Op-amp projects

Operational amplifiers (op-amps), such as the common type 741 (often sold as LM741), are packaged as ICs and provide a very easy way of amplifying voltages. Op-amps are often used on a 'dual rail' (sometimes called split rail) supply. This type of supply has already been mentioned, and is illustrated in Figure 5.1, along with an op-amp.

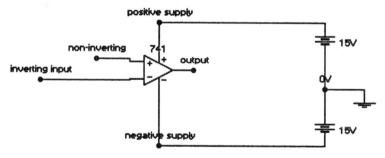

Figure 5.1

Figure. 5.1 shows how the dual rail power supply is created by using two batteries. They have been changed to 15 V batteries, as this is standard for op-amp circuits, but 12 V batteries will work equally well. Notice that the junction between the two batteries is called 0 V. We have ensured that Workbench knows this is 0 V by connecting the ground symbol. This is not always necessary in real life, although sensitive amplifier circuits are often connected to ground in the same way.

The *inputs* of the op-amp are labelled '+' and '–', like the power supply connections, but this does *not* mean positive and negative. When associated with the input connections, the + sign means 'non-inverting' input, and the – sign means 'inverting' input.

Testing a comparator circuit

Try the circuit shown in Figure 5.2. Note that two op-amps are provided. For now, select the type that provides power supply connections.

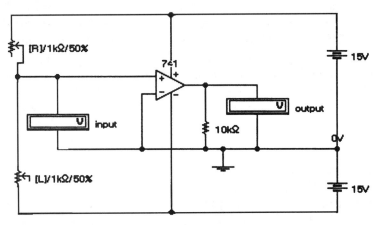

Figure 5.2

Note that one of the two inputs is tied to a known voltage. In this case, we have tied the inverting input to 0 V – this gives the circuit a reference point. The op-amp compares the voltages at its two inputs: if the non-inverting voltage is higher, the output becomes positive; if the inverting voltage is higher, the output becomes negative.

We can control the voltage at the non-inverting input by means of the two variable resistors. Double-click the lower one and change its KEY to L. The input voltmeter shows the voltage at the non inverting input. This can be changed by pressing L or R (with or without SHIFT).

The $10 \,\text{k}\Omega$ resistor is the load. In practice, this might be an LED (with series resistor), or a transistor circuit – more of this later. The output voltmeter shows the state of the output.

Experiment with the settings of the variable resistors, noting the readings on the voltmeters.

Results

You should notice that, if the voltage at the non-inverting input is made slightly higher than zero, the output voltage swings to nearly the positive supply voltage. If the input is made slightly less than zero, the output voltage swings to negative. Note that the output rarely hesitates – it is either positive or negative. Hence, the smallest change of voltage at the input can cause a very large swing at the output. Note also that the output does not move fully to +15 V or –15 V; an ideal op-amp would, but the fact that the 741 is not ideal is rarely important. The circuit is known as a 'Comparator', as it compares the voltages at the two inputs.

Try swapping the inverting and non-inverting inputs, i.e. connect the non-inverting input to 0 V and connect the inverting input to the junction between the two variable resistors. Your results should be the opposite of the previous ones.

A practical application

If the variable resistor with KEY L was replaced by a light-dependent resistor (LDR) and the other variable resistor raised to about $470 \,\text{k}\Omega$, the circuit could be used as an automatic light sensor. When the light striking the LDR is reduced, its resistance would rise, causing the input voltage to rise. If the variable resistor (with KEY R) had been carefully set, this rise of voltage could be made to trigger the output. The next circuit shows how the same idea can be achieved with a single rail supply.

Single rail comparator

This type of circuit can be operated with a normal battery.

Try the circuit shown in Figure 5.3. Note that the negative side of the battery is now connected to ground, and we will therefore

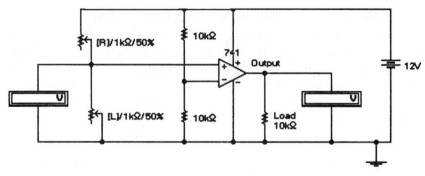

Figure 5.3

call this 0 V. If we connected the inverting input to 0 V, it would not be possible for the voltage at the non-inverting input to move below 0 V. Hence, the output could not be controlled properly. Therefore, we create an artificial mid-way voltage using a potential divider. As we saw in Chapter 1, a potential divider can simply be two resistors in series. If their values are equal, then the voltage where they join will be half the supply voltage. Now that we have tied the inverting input to 6 V (assuming a 12 V battery), the voltage at the non-inverting input can be moved below or above 6 V. Slight movements here will cause the output to swing between 0 V and 12 V.

Change the percentage of one or other of the variable resistors and note the readings on the meters. Does the output swing fully to 0 V and fully to 12 V?

Using the circuit

Like the previous circuit, one or other of the variable resistors could be an LDR (for light detection), or a thermistor (to detect temperature changes), etc. The output resistor labelled 'Load' would normally be something more useful than just a resistor. It could be an LED (with series resistor), but, although there is a considerable voltage swing at the output, the *current* available is too small to operate a buzzer, or even a relay. Hence, we need a device to amplify the current available at the output. The perfect current amplifier is the transistor, and this is employed in the next circuit.

A practical comparator

Try the circuit shown in Figure 5.4. Note that the power of the lamp has been changed to 1 W. In practice, the lamp could be an LED with series resistor (about 470 Ω to 680 Ω with a 12 V supply), or a buzzer, relay etc. The transistor could be BC108 or BC184, or any small npn type.

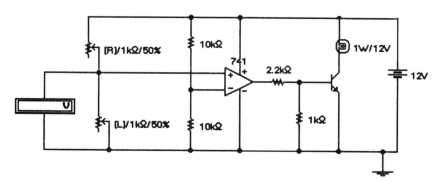

Figure 5.4

Note how the output from the op-amp is delivered to the transistor via another potential divider (if the 2.2 kΩ resistor is drawn vertically it will be more obvious). The purpose of this arrangement is to reduce the voltage at the base of the transistor. In the previous circuits the voltmeter at the output revealed that the voltage never falls much below about 0.9 V. The 741 seems incapable of making its output fall to 0 V. There are alternative ICs available that will overcome this problem, but a simple solution is to reduce the output by means of the potential divider arrangement, as shown.

One of the variable resistors could be a sensor, as described earlier. The other variable resistor allows the circuit to be set for the required effect, and should have a maximum value roughly equal to the resistance of the sensor at the point where a change of output is required. For example, if an automatic light is required, it is helpful to find out the resistance of the LDR at the point when you wish the lamp to light. If the resistance is, say, 200 kΩ then the variable resistor should have a maximum resistance of around 47 0kΩ. If in doubt, select a variable resistor with a higher maximum value, as it can always be reduced. However, if the value is too high it will be difficult to adjust the variable resistor accurately.

The maximum current that can be switched safely via the transistor depends upon the type selected. Those mentioned will tolerate over 100 mA, but avoid currents in excess of about 150 mA. If a larger current is required, then replace the transistor with a darlington pair, such as TIP121. The maximum current will now be over 3 A.

Selecting resistor values

Note: Skip this section if you prefer to avoid calculations and are content to find values by experimentation!

We have seen that, if the two resistors in a potential divider are equal, the voltage at their junction will be half the supply voltage. So does it matter whether the two values are say, 1 Ω, 1 kΩ, 100 kΩ, 10 MΩ, 5.6 kΩ, 47 kΩ, etc, etc?

The answer is yes and no... If we discount the two extremes in the list above, that is 1 Ω and 10 MΩ, any of the other values may well be satisfactory, both in Workbench and in real life.

However, our practical circuit based on Figure 5.4 should always be as energy efficient as possible – this will help our batteries to last longer. As there is a continuous flow of current through the two resistors, it will help if their values are as high as possible. So why not a billion ohms? The problem is that the op-amp is not perfect. It requires a small current at its inputs in order to determine the output voltage. As this current must be obtained from the positive supply rail via one of the two resistors, more current will be flowing through the upper resistor than the lower. The voltage difference across a resistor depends upon the current flowing. So, if the currents through the two resistors are unequal, the voltage across each resistor will be unequal. The net result will be a voltage at the junction between the resistors that is lower than the expected voltage.

The current required by the inputs of a 741 is quite small, but if the two variable resistors are changed to 1 MΩ values, and set at, say, 50 % each, the voltmeter will reveal the error, i.e. the voltage will be less than 6 V. There are op-amps available (e.g. with field effect transistor inputs) that require many times less current than a 741 at their inputs, but there will still be a small error.

Therefore, the resistor values must not be too low (or too much current will be wasted), nor too high, otherwise there will be a significant voltage error. We need a compromise based on a guide rule.

Rule of 10

In general, an error of less than 10% is acceptable (though it depends on the circumstance), but an error of more than 10% may cause problems and should be avoided. *Hence, if the current flowing through the pair of resistors is at least 10 times greater than the current required by the inputs of the op-amp, then all should be well.* A good data book or catalogue will provide information to obtain the input current. However, the 10 kΩ resistors suggested in Figure 5.4 will conduct more than 10 times the current required by the inputs of the worst op-amp, and are high enough in value not to waste current.

Variable resistor values

As one of the variable resistors in Figure 5.4 is likely to be a sensor in real life, the resistance of the sensor may already be set by the manufacturer. Therefore, all that the circuit designer has to do is to select a variable resistor that roughly matches the resistance of the sensor at the point where the circuit needs to switch. This was discussed above. Hopefully, the resistance of the sensor will not be so high or so low as to cause problems. Thermistors can be purchased in a variety of values (the value is normally stated at a temperature of around 25 °C). If a sensor had a very high value and the rule of 10 was compromised, it might be necessary to select an op-amp with a higher input impedance (such as an fet type).

Output resistors

We must ensure that the transistor will turn on and off correctly.

Turning on

The selected resistor values must not allow more current to flow than the output can supply. A 741 op-amp should not be expected to deliver much more than about 5 mA from its output. The values selected in Figure 5.4 maintain an output current of less than 5 mA. Check this by calculation, or by placing an ammeter in series with the output. The calculation is a little complicated, as current appears to be flowing through the 2.2 kΩ and 1 kΩ resistors in series, but the voltage at the base will be set by the transistor (say 0.86 V); hence it is only the 2.2 kΩ resistor that determines the current. The voltage at the output of the op-amp will be around 11.1 V.

Having established that the total resistance is acceptable, it must also be established that the gain of the transistor is sufficient to boost the current (via its collector) to the desired amount.

Turning off

Turning the transistor off is more difficult, as we have already seen that the voltage at the output of a 741 refuses to fall to the negative (or 0 V) supply rail. The ratio of the two resistor values is chosen to allow the voltage at the base of the transistor to be less than 0.4 V when the op-amp output is at say 1 V. A poor op-amp may not even manage to swing its output down to 1 V, so, if in doubt, the 2.2 kΩ resistor could be raised to 3.3 kΩ, or a silicon diode (e.g. 1N4148) could be placed in series with the output. As a diode causes a voltage drop of nearly 1 V, its inclusion will neatly solve the problem!

If a darlington pair transistor arrangement is used at the output there is less likely to be a problem, as a voltage of around 1.4 V is required to switch on a darlington pair – as described previously.

Inverting amplifier

The op-amp is capable of amplifying a continuously changing signal. To achieve this, we need to allow the output to swing smoothly between positive and negative, rather than jump between the two levels, as was the case with the comparator.

Negative feedback

We can make the output voltage move smoothly by means of 'negative feedback'. In other words we connect the output to the *inverting input*.

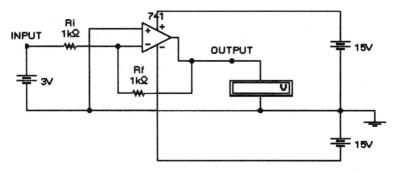

Figure 5.5

Try the circuit in Figure 5.5 (or, for a simpler faster simulation, try Figure 5.6). Note that we have returned to the dual rail 15 V supply. Making this type of circuit operate on a single rail supply is a little more complicated and will be covered later.

The 3 V battery supplies a DC signal. The voltmeter shows the resulting signal at the output. With the figures shown, the output will be −3 V. Try doubling the value of Rf to 2 kΩ − the output voltage will double to −6 V.

Cutting corners

Look at Figure 5.6: it is exactly the same as Figure 5.5, but we have used the op-amp symbol without its power supply connections. The dual supply is assumed by the simulator. The result is a simpler diagram and a much faster simulation. From now on we will design the circuits this way, but, in real life, the power supply connections must, of course, be included.

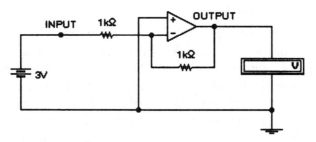

Figure 5.6

The default op-amp is the most efficient in terms of simulation speed, but note that the simulator assumes a power supply of ±20 V. However, this can be changed to ±15 V by double-clicking the op-amp, then clicking on EDIT and changing POSITIVE VOLTAGE SWING to 15 V and NEGATIVE VOLTAGE SWING to −15 V.

Results

The ratio of resistors Rf and Ri directly affect the voltage gain of the circuit in Figures 5.5 and 5.6. The output is inverted, i.e. there is a minus sign on the voltmeter, indicating a negative voltage. If the input voltage is made negative (by disconnecting, then rotating the 3 V battery), the output will be positive.

Voltage gain

Voltage gain (i.e. the amount of amplification) is defined as the ratio of the output voltage compared with the input, i.e.:

$$gain = output\ voltage/input\ voltage$$

We have found that the ratio of Rf to Ri sets the gain of the amplifier, hence:

$$gain = -(Rf/Ri)$$

We must not forget that the output is of the opposite polarity to the input. This is shown by the negative sign in the formula.

Further experiments

Try other values of input voltage by double-clicking on the 3 V battery. Try other values for Ri and Rf and check that the output voltage agrees with the formulae above. Is there any limit to this output voltage? Remember that no circuit of this type can provide an output greater than the supply voltage. In the circuit shown in Figure 5.5, the supply voltage is determined by the power supply batteries. However, in Figure 5.6 the simulator assumes a supply of ±20 V unless you have changed the op-amp parameters.

Amplifying alternating signals

The circuit in Figure 5.7 is the same as for Figure 5.6, except that a function generator is employed to supply an AC signal. In real life, this signal could be from a radio tuner or cassette player. The oscilloscope displays the result. The function generator is used to create a sine wave (an AC source could have been used instead).

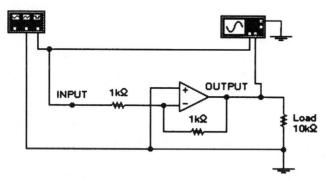

Figure 5.7

1 Drag the function generator from the top of the screen, double-click and set to 1 kHz (not the default of 1 Hz).
2 Set the amplitude of the function generator to 2 V.
3 Drag the oscilloscope into place. The oscilloscope ground connection has been joined to a ground symbol. This serves as a reminder that in real life the oscilloscope ground *must* be connected. However, this may be omitted in Workbench.
4 Double-click the oscilloscope and set its timebase to 0.20 ms/div.
5 Double-click the leads that connect to the oscilloscope and set them to different colours. This allows you to see which trace is which. Note that one trace shows the signal we are inputting to the op-amp; the other trace shows the output.
6 Activate the circuit.

Look carefully at the traces on the oscilloscope screen – when the input trace rises, the output trace falls. It is like a mirror image. Whatever the voltage at the point labelled 'input', the voltage at the 'output' will have the opposite polarity. This circuit is known as an 'inverting' amplifier.

We have already seen that the gain of the circuit is determined by the ratio of resistors Rf and Ri. Try changing the value of Rf to 2 kΩ and check that the output wave trace has twice the amplitude.

Beyond the limit

We appear to have the perfect system – an amplifier that can be made to increase our input signal voltage by any amount; all we need do is select a large enough value for Rf. Can we really fill the world's largest stadium with sound by employing our 741 op-amp with the appropriate resistors? Not quite! To begin with, the 741 can only deliver a current of a few milliamps – just enough to drive a pair of headphones – and the 'output' voltage is limited by our supply voltage.

Try increasing Rf to 15 kΩ. Look at the output trace on the oscilloscope. It will be easier to see the wave if channel B is changed to 10 V/div. You will see that the output wave is 'clipped'. In other words when it tries to move past the supply rail voltage it cannot, and instead becomes a straight line. The result of this would be severe distortion known as 'clipping distortion'. Most people will have experienced the sound that results when amplifiers, TVs, etc. are set to too high a volume.

Virtual earth

An op-amp with negative feedback maintains its output at exactly the right voltage to ensure that the voltage at its inverting input copies the voltage at its non-inverting input. This rule means that, if the non-inverting input is held at 0 V (by being connected to ground in Figure 5.7) then the inverting input will also remain at 0 V (or so close to 0 V that the difference can be ignored). Hence, the inverting input is known as a 'virtual earth'. This fact makes it a very useful mixer circuit, as described later.

Input impedance

Matching impedances between different circuits is quite a complex area and only a brief guide will be provided. Impedance can be thought of as the total resistance to an AC signal.

The input impedance of an amplifier determines how much current it requires at its input. The higher the input impedance, the less current required. The input signal flows through Ri to the inverting input, which, we have already seen, is a virtual earth. The signal will actually flow through Rf as well, but this is of no consequence, as the input impedance will be governed by the impedance experienced by the input signal on its way to ground.

Hence, the input impedance of the inverting amplifier in Figure 5.7 is simply the value of resistor Ri. This value must not be too large, since Rf has to be larger anyway to achieve a reasonable gain. In practice, Rf should not be greater than about 1 MΩ, and this often affects the choice for Ri.

However, if Ri is too low the amplifier will demand too much current from the input. Such a demand could cause the input voltage to collapse. For example, a crystal microphone has a very high impedance output. It can supply a reasonable voltage but very little current. If connected to the input of our amplifier, the 1 kΩ resistor (Ri) will allow so much current to flow that the microphone voltage will collapse to virtually zero.

So how do we select the value of Ri? The rule of 10 can be applied again: if the input impedance of our amplifier is 10 times (or more) higher than the output impedance of whatever is connected to input, then we will not affect the voltage of the input signal too much. However, this is a grey area – we often compromise, fudge, take a guess, and hope that all will be well.

In a real circuit, a value of 10 kΩ for Ri will generally prove satisfactory (apart from use with crystal microphones – where a different amplifier design is required).

Inverting amplifier with a single rail supply

The use of a dual rail supply allows an AC signal (e.g. from a microphone or tape recorder, etc.) to be amplified, with both the input and output waves moving above and below the 0 V rail. A dual rail supply can be obtained quite easily via a suitable mains-driven system, but for battery powered equipment it is much less convenient.

It is possible to make the op-amp work on a single rail supply. Try Figure 5.8, but note that the input signal must be at a low level, say 3 V. The two resistors R1 and R2 form a potential divider.

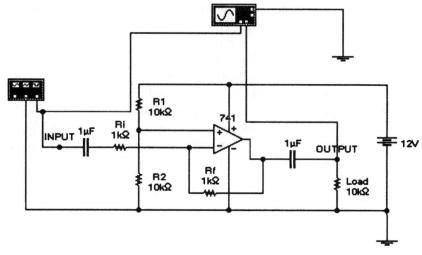

Figure 5.8

Assuming a 12 V supply, the voltage at the junction of R1 and R2 will be 6 V.

The op-amp rule that 'the output sits at the correct voltage to ensure that the voltage at the inverting input equals the voltage at the non-inverting input' means that the voltage at the inverting input in this case will be 6 V. In other words the average level of the input signal will be 6 V – the wave will move above and below 6 V. The capacitor at the input separates the average input signal (which is zero) from the 6 V average. The capacitor acts like a canal lock allowing the boats (AC signal) to pass, but separating the two water levels (DC levels).

The output also sits at half the supply voltage, and another capacitor is needed to drop the average DC level back to zero. Notice that the maximum amplitude of the output signal is much less than with the 15 V, 0, –15 V dual rail supply. In fact, with a 12 V battery, the output amplitude must be less than 6 V. Of course, the battery voltage may be raised to a maximum of 30 V, but this may be inconvenient in real life.

Summing amplifier

The summing amplifier is based on the inverting amplifier, and it may be driven from a dual rail or single rail supply. We will begin

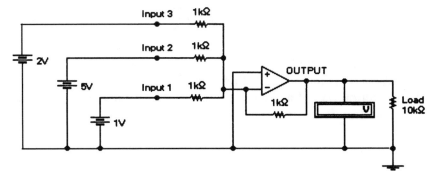

Figure 5.9

with a dual rail supply summing amplifier – which can be used to add DC voltages. We will then see how a similar circuit is used to make an audio mixer, but for variety this will driven from a single rail supply.

Try the circuit shown in Figure 5.9. Note that the op-amp is powered by a dual rail 15 V supply, but, as discussed earlier, it is common practice not to show the power supply. The three batteries are simply to create three different voltages to show how the circuit may be used to add or subtract.

1 Activate the circuit. The voltmeter shows the DC output level. Disregard the negative sign for the moment. Notice how the reading is the *sum* of the battery voltages.
2 Disconnect one of the batteries, and rotate it so that its negative side is uppermost. Reconnect it and activate the circuit once more. Notice how the reversed battery voltage is subtracted from the total.
3 Double-click on the various batteries and try a variety of voltages. Further batteries and 1 kΩ resistors may be added. Note, however, that the output cannot exceed the supply, which is generally 15 V.

Problem

In real life, it would be a nuisance to have a figure on the voltmeter that registered the opposite of the true figure. How would you make a circuit that provided a true figure with the correct polarity? **Note:** You will need a second op-amp.

Further work

At present, the values of the feedback resistor (Rf) and the input resistor (Ri) are equal. If the ratio of values is changed, the output will reflect this change according to the formula provided under the earlier section on voltage gain. Hence, if Rf is changed to 2 kΩ, the output will be twice the previous value. Try it and see.

Audio mixer

The summing amplifier can be used to add AC signals and it thus makes an ideal audio mixer. For example, you may wish to add commentary and music to a video soundtrack. An amplified signal from your microphone could be added to the signal from a tape recorder to produce a mixed output. A suggested circuit is provided in Figure 5.10. A dual rail supply could be employed, but a single rail circuit is little more trouble when mixing AC signals, and so the single rail version is shown.

Try the circuit. Notice that, as there is only one signal generator available, we have employed two AC voltage sources. These can be independently set regarding voltage and frequency. Double-click and set them as indicated. The oscilloscope will show the resulting mixed signal.

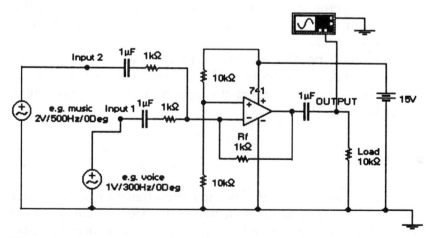

Figure 5.10

Any number of sources can be added in the same way. Note that, as the circuit is operating on a single rail supply, each source requires a capacitor in addition to the resistor. In this example, we have used 1 kΩ resistors. However, in practice, the use of 10 kΩ resistors for both the feedback and input will provide a higher, and therefore more useful, input impedance.

Controlling the input signals

The circuit in Figure 5.11 shows how potentiometers may be employed to control the input signal levels. Note that the feedback and input resistors have been changed to 10 kΩ as suggested above, so that the rule of 10 is obeyed, i.e. the input impedance of the inverting amplifier is 10 times more than the value of the potentiometer.

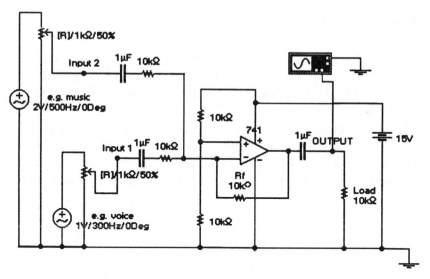

Figure 5.11

In a practical circuit, the feedback and input resistors could be higher still, say 47 kΩ, allowing the use of higher potentiometer values. This is an advantage as less current will be demanded from the sources (tape recorder, etc.) feeding the mixer.

Our mixer is now in the form of a circuit that has a clear practical use. The input voltage levels should be around 1 V (remember we are dealing with AC signals), in which case the output will be around the same value. Most power amplifiers require 'line input signals' of around 1 V. Line signals are the type provided by cassette recorders, CD players, etc. Note that record decks do not normally include a pre-amp and require a special input to the power amplifier known as the 'phono' input. This both amplifies and applies frequency correction to the signal. Microphones produce very small signals in the order of a few mV. Before they can be applied to the mixer they must be amplified by a pre-amp. The source labelled 'voice' in Figure 5.11 assumes that the microphone signal has already been amplified.

The fact that the output is out of phase compared with the input (i.e. when the input goes positive, the output goes negative) is of no consequence in amplifiers or mixers, as our ears (fortunately) cannot detect phase differences.

Non-inverting amplifier

It is possible to apply a signal to the non-inverting input, but note that the circuit and calculations differ from those for the inverting amplifier. Figure 5.12 shows how the non-inverting amplifier is arranged.

Try the circuit, noting again that the battery is *not* the supply, but is a voltage applied to the input. The output is displayed on the voltmeter. Using the figures in the diagram, a reading of 6 V will be obtained on the voltmeter. Notice that the reading is positive. This

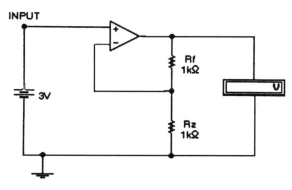

Figure 5.12

is because we are using the non-inverting input, hence our output will have the same polarity as the input.

Voltage gain

The gain of the circuit is 2, assuming the figures in Figure 5.12. Resistor Rf is still the feedback resistor – note that it is effectively connected from the output to the inverting input. The resistor labelled Rz is connected between the inverting input and ground.

The gain, which is still the ratio of output voltage to input voltage, is given by:

$$\text{gain} = (\text{Rf/Rz}) + 1$$

So, in Figure 5.12, where the two resistors have a value of $1\,\text{k}\Omega$, the gain will be 2.

Try making Rf equal to $2\,\text{k}\Omega$. The gain should be:

$$(2\,\text{k}\Omega/1\,\text{k}\Omega) + 1 = 3$$

If the input is a negative voltage (which can be achieved by disconnecting the battery and rotating, then reconnecting it), the output will be negative. Experiment!

Amplifying AC signals

Try the circuit in Figure 5.13. Set the frequency of the function generator to $1\,\text{kHz}$, and its amplitude to $2\,\text{V}$. Set the timebase of the oscilloscope to $0.5\,\text{ms/div}$. As usual, the oscilloscope traces will be clearer if the wires connecting the oscilloscope are coloured (by double-clicking).

You should notice that the output trace is twice the amplitude of the input (assuming the values in Figure 5.13) and is in phase. In other words, when the input trace moves up, the output moves up as well.

Input impedance

The input impedance of the dual rail supply version (i.e. Figure 5.13) is equal to the input impedance of the op-amp itself. This is

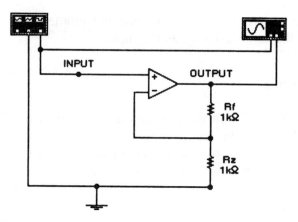

Figure 5.13

typically very high, and, with some op-amps, so high it can be considered virtually infinite. This is a considerable advantage over the inverting amplifier shown earlier.

The disadvantage is that the input signals are not delivered into a virtual earth like the inverting circuit, so the non-inverting circuit is not useful as a mixer. The main application of the non-inverting circuit is as a high impedance amplifier. It works well as a microphone amplifier, where a high input impedance is useful.

Non-inverting single rail amplifier

The non-inverting amplifier can be made to operate on a single rail power supply. The advantage is convenience in being able to use a single normal battery or inexpensive power unit: the disadvantage is a reduced input impedance and the need for three capacitors.

Try the circuit in Figure 5.14. Set the function generator to a frequency of 1 kHz, and amplitude of 2 V. Set the oscilloscope timebase to 0.5 ms/div. As usual, the colours of the leads connecting the oscilloscope can be changed by double-clicking; this improves the display considerably.

Note that the input and output signals move above and below zero. However, if the oscilloscope leads are moved to the non-inverting input, or the op-amp output (i.e. the op-amp side of the capacitor) then the wave traces will be shifted up by 6 V (assuming a 12 V supply).

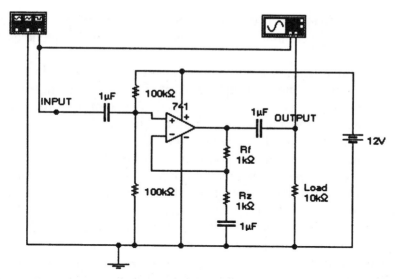

Figure 5.14

The two $100\,k\Omega$ resistors hold the non-inverting input at an average of 6 V. Remember the rule that an op-amp (with negative feedback) must make the output whatever voltage it has to be to cause the voltage at the inverting input to equal the non-inverting input. This means that the voltage at the inverting input must be maintained at 6 V. As virtually no DC flows through the feedback resistor, the output will also remain at an average of 6 V. The three capacitors ensure that no DC that would upset the conditions described will flow.

When our AC input signal (from a microphone, for example) passes through the capacitor it makes the non-inverting input wobble up and down on either side of 6 V. A similar signal appears at the output of the op-amp. This signal may be amplified according to the formula mentioned earlier, namely:

$$gain = (Rf/Rz) + 1$$

Selecting the capacitors

Capacitors conduct AC and block DC – this is why they are so useful in this circuit. However, capacitors do not conduct all frequencies equally well. Higher frequencies will be conducted more easily than lower frequencies. So the capacitors in this circuit must have large enough values to conduct the lowest frequencies of interest.

The human hearing range is from about 20 Hz to 20 kHz. If our capacitors will conduct frequencies of 20 Hz without noticeable loss, then all will be well. In practice, the situation is complicated by the fact that the amount of current required will affect the value of capacitor needed. This is where trial and error using Electronics Workbench is so valuable. You can try different capacitor values and check the results on the oscilloscope. The values of all the resistors will affect the size of the capacitors required. As a very rough guide, and with the values of resistors given in Figure 5.14, an input capacitor value of 0.47 µF (470 nF) is sufficient. Values of 10 µF should be sufficient for the other capacitors.

In real life, the input capacitor should be a good quality type, such as a 'polylayer capacitor', and the larger capacitors must be 'electrolytic' types. Electrolytic capacitors are polarized and must be connected the correct way round with respect to positive and negative. In both cases, the positive side must face towards the op-amp.

Input impedance

The input impedance of the dual rail power supply circuit (Figure 5.13) is very high – effectively the input impedance of the chip itself. However, the single rail supply circuit includes the two 100 kΩ resistors maintaining the non-inverting input at half the supply voltage. The input signal will flow through *both* these resistors (even the one connected to positive) making the effective input impedance the value of the two 100 kΩ resistors in *parallel*. So, the effective input impedance will be 50 kΩ.

The 100 kΩ resistors could be raised in value if a higher input impedance is required, but remember that any current flowing into the op-amp will upset the voltage at the non-inverting input and cause an error. The higher the resistor values, the greater the error. It is possible to check the voltage using Workbench, but note that the voltmeters employed by Workbench have a resistance of 1 MΩ, and the current flowing through the voltmeter will add to the error! The multimeter at the top left hand corner of the screen can be used instead of the standard voltmeter; the internal resistance of the multimeter can be set to 1 TΩ, i.e. $1 \times 10^{12} \, \Omega$!

Filters

The chapter would not be complete without a look at how we can filter certain frequencies by means of an op-amp. Most hi-fi amplifiers have bass and treble controls: cheaper amplifiers may simply have a 'tone' control. All these controls work with circuits that can boost or cut certain frequencies. We will begin by looking at the principle of using a capacitor and resistor network.

Treble cut filter

1 Try the circuit shown in Figure 5.15. Notice that it consists of only two components: a resistor and a capacitor.

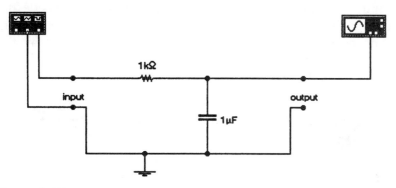

Figure 5.15

2 At the input side connect the function generator and set its amplitude to 10 V and its frequency to 1 Hz.
3 Set the oscilloscope timebase to 0.20 s/div. Note that the oscilloscope need not be connected to ground – though in real life it must be. However, the circuit must be referenced to ground as shown.
4 Activate the circuit and note that the graph on the oscilloscope screen is a faithful copy of the wave produced by the function generator. The spare input of the oscilloscope could be connected directly to the dot above 'input' to prove this, if required.
5 Now change the frequency of the function generator to 1 kHz.
6 Change the timebase of the oscilloscope to 0.20 ms/div (i.e. 1000 times faster). Note that the signal passing through the circuit is

now attenuated (reduced in amplitude). This is because capacitors block DC but conduct AC. The higher the AC frequency the more easily it is conducted through a capacitor. As the capacitor in Figure 5.15 is connected from the output side to 0 V, the higher frequencies are 'shorted' to 0 V.

It is possible to plot a graph of the effect of the resistor/ capacitor on the amplitude of the wave at various frequencies. This is tedious and, if you wished to compare the effect of different resistor/capacitor values, it would be *very* tedious! There are some mathematics that can help, but Workbench provides a miraculous device (which does not exist in real life) called a bode plotter.

The bode plotter calculates the effect of the circuit on a range of frequencies and plots a graph that compares frequency with amplitude. Although the bode plotter does not exist outside the computer, the circuit under test can be constructed with reasonable confidence that it will perform in real life as predicted by the computer.

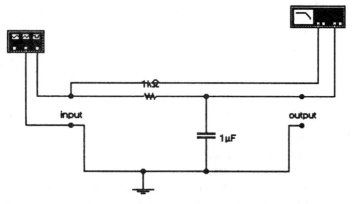

Figure 5.16

The bode plotter is shown in Figure 5.16. Like the oscilloscope, it does not require a ground connection, but both its input and output must be connected as shown. The bode plotter also requires the function generator to create the sine wave signal needed for plotting. The settings on the function generator are controlled by the bode plotter.

1 Double-click the bode plotter and check that MAGNITUDE is selected.
2 Both the horizontal and vertical scales should be set to LOG.
3 The frequency range should be set to, say, 1 Hz – 10 MHz. This is achieved by changing the settings in the boxes below the word HORIZONTAL. The bode plotter is now ready for action.

Activate the circuit – you will see the frequencies being scanned. Slower computers may take some time to achieve a plot. At the end of scanning, a graph will be drawn. The graph shows that the higher the frequency (horizontal axis), the lower the amplitude (vertical axis). The frequency is measured in Hz (or kHz, or MHz) and the amplitude is measured in decibels (dB).

As the signal is neither amplified nor attenuated to begin with, the bode plotter provides a reading of 0 dB at the start. If the circuit amplifies the signal, the output will have a positive reading; if the circuit attenuates the reading, a negative dB level will result.

The bode plotter allows you to check the dB level at any point by moving a vertical line horizontally along the scale. At present, the vertical line will be at the left hand side of the screen. Use the mouse to click on the bottom of the line, and move it to the right. As it moves, two numbers will change on the bode plotter; these show the dB level and the frequency of the plot at any one stage. At the left-hand side, the dB level will be zero, showing that the signal has not been changed. As the line is moved to the right, the dB level will fall, showing the amount of attenuation at any particular frequency.

The point at which the plot suddenly begins moving downwards is known as the break frequency. The break frequency can be calculated as follows:

$$\text{break frequency} = 1/(2\pi RC)$$

Remember to measure R in ohms and C in farads. However, the bode plotter avoids the need for calculation by displaying the break frequency; simply move the vertical line to just beyond the corner of the line (the dB reading will be about –3 dB). The frequency reading will be the break frequency.

Bass cut filter

The circuit shown in Figure 5.17 illustrates that, by swapping the resistor and capacitor around, we can make a filter that reduces the

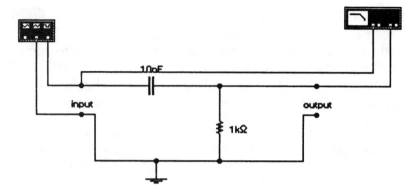

Figure 5.17

amplitude of the wave at lower frequencies. The capacitor, as before, allows higher frequencies to pass with less attenuation than lower frequencies. As the capacitor is in series, the higher frequencies are able to pass easily from the 'input' to the 'output'.

Try the circuit, setting up the bode plotter as before. The graph should show the amplitude rising and then levelling off at a frequency of about 20 kHz. Try changing the value of the resistor, or capacitor, or both, and replotting. The break frequency is calculated using the same formula as before.

The filters shown so far are known as 'passive filters', as they can only reduce the amplitude of a signal. Their behaviour depends upon the amount of signal used from the output – our bode plotter used very little – but this might not be the case in real life. We will now see how matters can be improved.

Active treble cut filter

Try the circuit shown in Figure 5.18. At low frequencies, the capacitor has little effect and so the circuit is an inverting amplifier with input resistor 10 kΩ and feedback resistor 100 kΩ. This provides a gain of 10. The output is out of phase with the input, but we will ignore the phase from now on, as the situation is complicated when capacitors are involved – the bode plotter will show phase differences if required. A gain of 10 is the same as 20 dB, hence the bode plotter indicates an initial gain of 20 dB. At

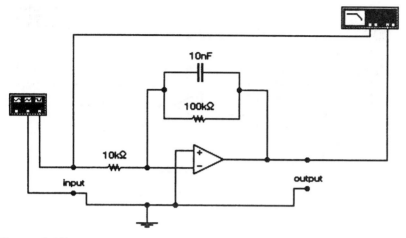

Figure 5.18

the break frequency (calculated as before, using the values of the capacitor and 100 kΩ resistor) the line begins to slope downwards.

The reduction in amplitude at higher frequencies is due, again, to the way in which a capacitor conducts higher frequencies more easily. As the capacitor is in the negative feedback loop, the *more* feedback, the *less* gain. Try other values of capacitor and/or resistor and observe the effect.

Active treble boost filter

Fig 5.19 shows how we can boost higher frequencies. Ignoring the effect of the capacitor, the circuit provides a gain of unity, i.e. no gain (0 dB). As the frequency at the input is increased, a larger signal passes via the capacitor, making the input resistance appear less. Hence, the gain of the circuit is increased. The break frequency is determined by the values of the capacitor and resistor in parallel.

There is a problem with this circuit – which may not be apparent until the real version is constructed. The graph on the bode plotter slopes upwards as the frequency increases and there appears to be no limit to the gain provided. We have limited the upper frequency displayed to 10 MHz, but, in real life, there may be no limit. In practice, the design of the op-amp may limit the upper frequency, but it is still bad practice not to design an upper limit.

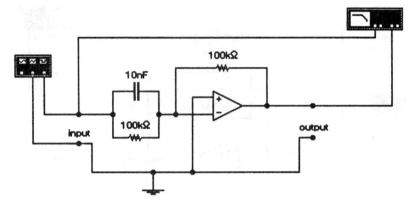

Figure 5.19

The circuit shown in Figure 5.20 overcomes this problem by including an extra series resistor. Even at the highest frequencies (when the capacitor appears to have zero resistance), there will always be an input resistance of $100\,\Omega$. Hence, the gain cannot rise above $100\,k/100 = 1000$ times ($60\,dB$)

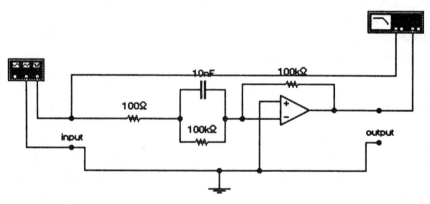

Figure 5.20

Active bass cut filter

The circuit in Figure 5.21 shows how an op-amp is employed to attenuate bass frequencies. Try the circuit. Working backwards – at high frequencies the capacitor will offer virtually no resistance to the signal, so we effectively have an amplifier with a $10\,k\Omega$

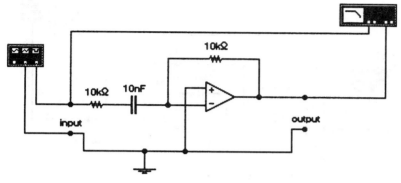

Figure 5.21

feedback resistor and $10\,k\Omega$ input resistor. Hence the gain will be unity ($0\,dB$). But at low frequencies, the capacitor will impede the input signal, so reducing the gain. The break frequency is determined by the values of the capacitor and resistor in series.

Active bass boost filter

Figure 5.22 shows how we can boost low frequencies. Again, starting at the top, the highest frequencies will pass through the capacitor as if it were a perfect conductor. The gain will be provided by dividing the value of the feedback resistor by the input resistor, i.e. unity gain or $0\,dB$. At lower frequencies, the capacitor will

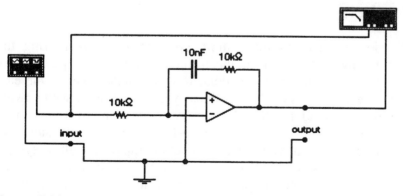

Figure 5.22

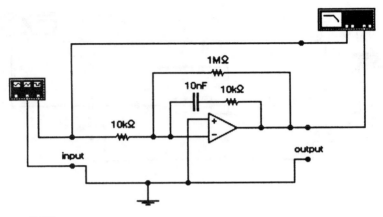

Figure 5.23

impede the signal, increasing the feedback value and hence increasing the gain.

Like the active treble boost filter, there is a problem with this circuit and it would almost certainly fail in real life. There must always be a DC feedback path to the inverting input of an op-amp, otherwise it will behave like a comparator, and its output will swing towards +15 V or −15 V. The capacitor in the feedback path blocks DC, so an additional resistor is required to provide a DC route.

Figure 5.23 shows the resulting circuit. The 1 MΩ feedback resistor provides a DC path regardless of the effect of the capacitor. This ensures that the gain – even at the lowest frequencies – cannot exceed 1 M/10 k = 100 (i.e. 40 dB).

Decibels

The decibel scale is often more convenient for measuring the gain or loss of signal in electronic circuits. It is particularly useful for measurements associated with sound levels. Our ears do not respond to sounds in a linear way. If sound A seems twice as loud as sound B, then A actually requires many times more generating power. This is why a hi-fi system rated at, say, 3 W seems louder than you might expect, compared with a system rated at 10 W. When you replace the 10 W system with one rated at 60 W, you may be disappointed that it does not sound six times louder!

The decibel scale is more in line with actual human hearing: 40 dB should sound twice as loud as 20 dB. The decibel scale is known as a 'log' scale – this is why the word log is indicated on the bode plotter. Sound levels are normally measured in a relative way – so we do not normally call silence 0 dB; instead, we use a reference point that is easier to measure. For example, the sound level indicators on a tape recorder or video recorder will show 0 dB as the maximum level permitted. Signals louder than this will have a positive dB reading; quieter signals will have a negative reading.

You may have wondered how to convert a voltage gain into a dB reading. The formula that links the two is:

$$\text{gain} = 20 \log (\text{V out/V in})$$

The 'log' in the formula is log to base 10. The gain will be a reading in dB. For example, if the ratio of Vout to Vin is say 1 (unity) then the gain will be 0 dB. If the ratio is 2, the gain will be 6 dB. A 10 times ratio will be 20 dB.

The same formula can be used to compare current ratios and provide a dB reading. However, *power* requires a different formula. In this case:

$$\text{gain} = 10 \log (\text{P out/P in})$$

Again, the log is to base 10, and the gain will be in dB. So, a power ratio of 2 will be equivalent to a gain of 3 dB.

When drawing graphs, the use of log scales (e.g. dB to measure gain) enables the whole story to be expressed within a reasonable area of paper. You will have noticed that the bode plotter also plotted frequency on a log scale. This allowed us to view frequencies from 1 Hz to 10 MHz without running out of space, but with the lower frequencies clearly defined. So 1 Hz to 10 Hz occupied the same length as 10 Hz to 100 Hz, 100 Hz to 1000 Hz, etc.

Questions

Full written answers, complete with Workbench circuits, are available on the accompanying disk. See p. 224 for details.

1 Design a comparator circuit based on an op-amp and a single rail (normal 12 V battery) power supply. Use three resistors each of

$10\,k\Omega$ and one variable resistor of $22\,k\Omega$. Connect a voltmeter between the output and $0\,V$ and show how the reading on the voltmeter can be made to switch between $0\,V$ and positive when the variable resistor is changed slightly.

2 Add a darlington pair transistor arrangement to light a $10\,W$, $12\,V$ lamp. Calculate the series resistor required at the base of the darlington pair, assuming that the op-amp output can achieve a maximum of $12\,V$ and that you must not draw more than $4\,mA$ from the output.

3 Draw an inverting op-amp circuit with an input resistor of $10\,k\Omega$ and a feedback resistor of $22\,k\Omega$. State:

   (a) the gain of the circuit
   (b) the output voltage if the input is connected to a $3\,V$ battery
   (c) the output when the input is a signal of $4\,V$ AC
   (d) draw two sketch graphs to show how the input in (c) compares with the output
   (e) now draw a third graph to show the shape of the output if the input signal is raised to $12\,V$ AC.

4 Design an inverting amplifier that increases a signal by 10 times with an input impedance of $47\,k\Omega$. If a signal of peak $0.2\,V$ is applied to the input, state the peak value of the output signal.

5 A summing amplifier (like the circuit in Figure 5.9) has two input resistors each of $10\,k\Omega$, and a feedback resistor of $33\,k\Omega$. Find the output voltage if $2\,V$ is applied to input 1 and $-3\,V$ is applied to input 2.

6 The same summing circuit as in Q. 5 is used, but the $2\,V$ input is applied to an input resistor of $5\,k\Omega$. The $-3\,V$ input is applied to a $10\,k\Omega$ resistor as before, and the feedback resistor is still $33\,k\Omega$. Find the new output voltage.

7 Design a summing amplifier with four input resistors labelled A, B, C, D. The values are: A= $8\,k\Omega$, B= $4\,k\Omega$, C= $2\,k\Omega$ and D= $1\,k\Omega$. The feedback resistor value is $1.6\,k\Omega$. Each input should be connected to a $5\,V$ battery by means of a switch. The four switches should be labelled A, B, C, D. Connect a voltmeter to the output.

   (a) find the output voltage when only A and C are switched on.
   (b) find the output voltage when only B, C and D are switched on.

(c) how could the circuit be used to convert from binary to decimal?

8 The circuit in Q. 7 produces negative answers. Add another op-amp circuit so that the reading on the voltmeter is positive.

9 Design a summing amplifier with two 10 kΩ input resistors and a 10 kΩ feedback resistor. Connect an AC voltage source to one input, and the function generator to the other. Set the AC voltage source to 1 V, 60 Hz. Set the function generator to 10 V, 1 Hz and click the 'triangular wave', i.e. the one in the centre. Connect an oscilloscope to the output and sets its timebase to 0.05 s/div. and its input to 5 V/div. Sketch a graph of the expected result and, if possible, compare your result with the trace drawn by Workbench.

10 Draw the circuit diagram of a non-inverting amplifier based on an op-amp. Select a 10 kΩ feedback resistor and make the value of the other resistor 5 kΩ. Calculate:

(a) the gain of the amplifier
(b) the output voltage if 4V DC is applied to the input
(c) the output if an AC signal of 2 V RMS is applied to the input
(d) the output if an AC signal of 9 V RMS is applied to the input.

11 An active treble cut filter was designed in an experimental attempt to reduce the noises produced by a badly scratched record. It was decided to try a break frequency of 10 kHz, i.e. sounds higher than 10 kHz would be attenuated (reduced). Below 10 kHz the gain should be unity (i.e. the output and input have equal amplitudes).

(a) design a suitable circuit
(b) calculate the resistors required to produce a break frequency of 10 kHz., assuming that the capacitor value chosen is 10 nF
(c) sketch a 'bode plot' graph to show the action of your circuit.

12 The same record was also warped, and produced an irritating low frequency rumbling noise. Design a bass cut circuit, with a break frequency of 100 Hz and a gain of 5 above this frequency. Assume that a capacitor value of 220 nF is employed.

(a) draw the required circuit

    (b) calculate the resistor required for a break frequency of 100 Hz

    (c) calculate the value of the resistor required for the gain of 5 as described

    (d) sketch a 'bode plot' graph to show the action of your circuit.

Chapter 6

Further op-amp circuits

Non-inverting Schmitt trigger

A Schmitt trigger is a circuit with positive feedback. In other words, part of the output is returned to the non-inverting input. This has the effect of making the input appear less sensitive – it requires a large voltage swing to change the output. An example of where this is useful is an automatic curtain winder. As the level of daylight falls, the circuit causes the curtains to close. It would be very irritating if slight changes in the daylight caused the curtains to open again. With a Schmitt trigger employed, the daylight has to rise by a large amount before the curtains open.

Schmitt triggers can be made in a number of ways, and a logic gate (see Chapter 7) may be employed. Looking at op-amp versions, we will see how these triggers can be extended to make waveform generators.

Try the circuit in Figure 6.1. The op-amp with dual rail supply has been selected as this is simulated much more quickly. However, we still need a dual rail power supply to create the input voltages, hence the 12 V batteries shown. It is essential to use a ground connection on the battery supply, as well as on the op-amp circuit.

When the circuit is first activated, the output may not rise or fall to the expected voltage. Change the setting of the potentiometer. The output voltage should change to plus or minus 20 V. When the potentiometer setting is changed the other way, the output should remain at a steady voltage until it swings fully in the opposite direction.

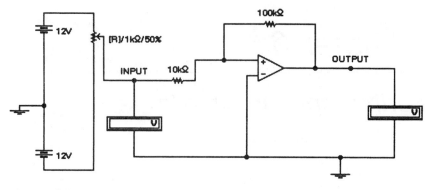

Figure 6.1

Try changing the potentiometer by pressing R or SHIFT R. The result of this action will be shown by the input voltmeter. The output voltage should rise to +20 V or fall to –20 V. The output should appear sluggish: this is the Schmitt effect.

Try changing the feedback resistor to 1 MΩ. The positive feedback effect will be much reduced, and so the output voltage can be toggled much more easily. The action of the Schmitt trigger can be demonstrated in a more automated way, as shown in Chapter 7.

Inverting Schmitt trigger

Try the circuit shown in Figure 6.2. Note that the batteries have been increased to 20 V. This is to match the default voltage assumed for the op-amp. The op-amp saturation voltage can be changed by double-clicking, then selecting EDIT. The positive voltage swing and negative voltage swing can be set as required.

When first activating the circuit, the output voltage will not saturate (i.e. move fully positive or negative) until the input voltage is changed (by pressing R or SHIFT R) sufficiently. Once the output saturates, the Schmitt effect will cause the output to switch cleanly between positive and negative.

The junction between the two 10 kΩ resistors will be at half the output voltage (check with a voltmeter). This means that the switching thresholds (i.e. the points at which the input voltage cause the output to switch) will be the same voltage, but with the opposite polarity. The switching voltages can be modified by changing the values of the two 10 kΩ resistors.

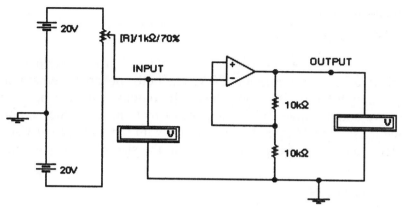

Figure 6.2

Relaxation oscillator

The circuit shown in Figure 6.3 is based on the inverting Schmitt circuit. Try the circuit, but note that, to begin with, the oscilloscope will display a straight line. This happens in real life, but as the oscillations occur in less than a second the delay is not noticeable.

The relaxation oscillator is useful, in that it always begins to oscillate. Some oscillators (or astables) require a pulse before they will begin to oscillate.

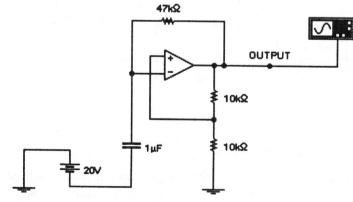

Figure 6.3

Applications

Flashing light

Try connecting a bulb between the output and ground. Remember to change its voltage rating to 20 V. The bulb appears to remain on, even though the oscilloscope shows the output switching between positive and negative. Why? Remember that bulbs conduct in both directions. Try adding a diode in series with the bulb. *Note:* Users of fast computers will need to increase the size of the capacitor.

Flashing LEDS

Connect an LED (and series resistor) between the output and ground. The LED should work, but in real life it will be damaged, as more than 5 V will be applied across the LED in the reverse direction each time the output goes negative (real LEDs cannot tolerate more than 5 V in reverse). The solution is to connect another LED in reverse parallel as shown in Figure 6.4. The second LED could be a different colour. If a second LED is not required, then use an ordinary diode instead.

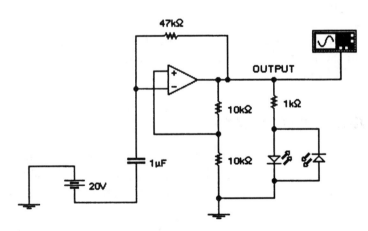

Figure 6.4

Lorry reversing warning

Connect a buzzer between the output and ground. Adjust the frequency of the buzzer and the values of the resistors, so that it sounds like the warning beepers fitted to buses and lorries.

Ramp generator

A ramp generator is sometimes called an integrator. Try the circuit in Figure 6.5. Note that the battery has been rotated, so that its negative end is connected to the resistor. When the switch is closed, the capacitor is discharged and the oscilloscope shows the output switching to 0 V. When the switch is opened the output voltage rises in uniform fashion. Set the oscilloscope timebase to 0.05 s/div to observe the trace. This circuit is very useful if a voltage is required to rise at a constant rate.

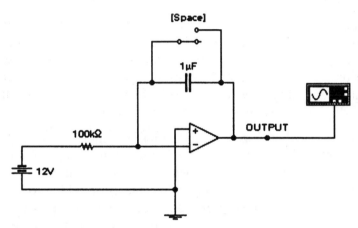

Figure 6.5

Sawtooth generator

If an oscillator is combined with a ramp generator a circuit known as a sawtooth generator can be produced. Try Figure 6.6. Notice that it is basically a combination of the circuits in Figs 6.4 and 6.5.

There will be a pause before the oscillator begins working. This pause will last for less than one second of *real* time, but may appear longer when using computer simulations. After the pause, a sawtooth wave will appear on the oscilloscope.

Sawtooth waves are used in TVs (and oscilloscopes) to make the electron beam scan across the cathode ray tube. The beam must scan smoothly across the screen, then move quickly back to begin the next line.

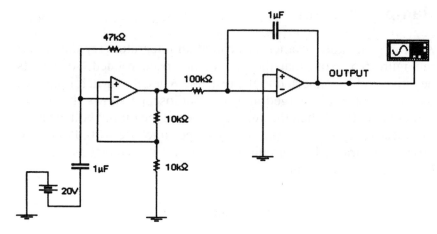

Figure 6.6

Voltage regulator

We have seen that a zener diode can be employed to regulate voltage. When such a diode is coupled to a transistor, it is possible to obtain a fixed voltage with a useful amount of current. The only problem is the voltage drop across the transistor, which lowers the output. If an op-amp with negative feedback is employed, the output voltage will be exactly the same as the zener voltage.

Try the circuit in Figure 6.7. Notice that the voltage across the zener diode is copied exactly across the output. The negative feedback applied to the op-amp is taken from the emitter of the

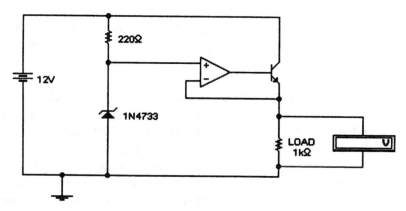

Figure 6.7

transistor – this ensures that the voltage drop across the transistor is allowed for. Remember the rule for op-amps with negative feedback: the output (from the op-amp) will be at whatever voltage necessary to make the inverting input voltage equal to the non-inverting input. Hence, in this circuit the output from the op-amp will be higher than 5 V in order to compensate for the voltage across the transistor. Check the output voltage from the op-amp with a voltmeter; it should be about 5 V.

Dual rail regulated voltage supply

Many of the op-amp circuits described require a dual rail power supply, i.e. a supply with positive, zero and negative rails. We will now see how such a supply can be created from a mains input. Look at the circuit in Figure 6.8. We have assumed a supply of 230 V AC and have employed a 10:1 transformer. Begin by double-clicking the transformer, selecting EDIT, and changing the turns ratio to 8:1. Note that the caption in the diagram will continue to say 10 to 1.

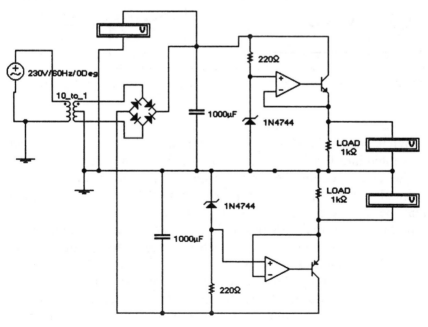

Figure 6.8

The 8:1 ratio will provide an AC output of 28.75 V. When rectified, there will be a loss of about 1.4 V leaving 27.35 V split equally across the circuit. The positive and negative sides of the circuit will have a PD of 13.7 V, but this will rise to 19 V (13.7 × 1.4) as the PD across each 1000 μF capacitor rises to the peak value. The op-amp regulators work in a similar way to the previous circuit, providing an output of ±15 V.

Difference amplifier

A circuit that amplifies the difference in voltage across its inputs is shown in Figure 6.9. Test the circuit, setting the AC voltage source (which represents an input transducer, such as a microphone) to 1 V, 400 Hz, and the oscilloscope to 1 ms/div. The wave trace should reveal an amplified copy of the signal, much like a normal amplifier. The 50 Ω resistors simulate the impedance of the leads.

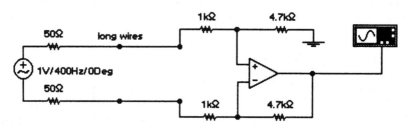

Figure 6.9

What makes this amplifier different is its ability to reject unwanted interference (electrical noise), which may be picked up in the wires connecting the input transducer to the amplifier. This is a particular problem with microphones, where the microphone signal may be as low as 50 mV. Microphone cables are often long and might pass near mains cables, lighting cables, stage dimming units, etc. It is not unusual for electrical noise to obliterate the microphone signal altogether.

Figure 6.10 shows how we can inject some unwanted noise, to see how our amplifier fares. In spite of the extreme conditions, where 120 V AC is connected to our input lines, the wave trace remains a perfect copy of *only* the wanted 400 Hz signal. The 1 kΩ

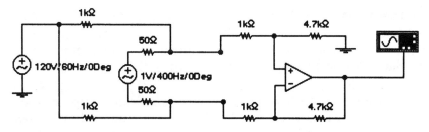

Figure 6.10

resistors in series with our noise generator are needed in the simulation because mains cables are never (we hope) connected directly to microphone leads. Instead, the mains interference is induced via magnetic and/or electrostatic fields.

It is important that *both* wires connecting the transducer with the amplifier pick up the same unwanted noise. If more noise is induced into one wire than the other, then some of the noise will find its way through the system. We can simulate this by disconnecting a wire from one of the 1 kΩ 'noise' resistors. The whole 400 Hz wave trace now wobbles up and down in phase with the 60 Hz noise. This would result in 'mains hum' being heard through the speaker system.

In practice, the wires linking the transducer with the amplifier are twisted together, to help ensure that they both pick up identical interference. Links such as this are often used to connect very high quality sound equipment. The twisted pair of cables are also screened with a mesh of wires connected to 0 V – this provides additional protection against unwanted noise.

Cheap microphones are designed to transmit a signal down a single wire, which is surrounded by a screen. The microphone element has two connections: one is connected to 0 V (e.g. the screen) and the other provides the signal. A normal amplifier, such as the non-inverting amplifier described in Chapter 5, might then be used to raise the signal (e.g. 50 mV amplitude) from the microphone to a 'line level' signal with an amplitude of about 1 V.

More expensive microphones are often 'balanced'. This means that the two connections of the microphone element are connected to the twisted pair described above. The 0 V screen is connected only to the metal case of the microphone. Balanced microphones are designed for use with a difference amplifier.

Questions

Full written answers, complete with workbench circuits, are available on the accompanying disk. See p. 224 for details.

1 Draw a circuit diagram of a non-inverting Schmitt trigger based on Figure 6.1. Use a 47 kΩ feedback resistor and a 10 kΩ input resistor. Check that the circuit works, and that it is possible to toggle the output between positive and negative. State the voltage required either side of zero, at the input, that causes the output to change state.

2 Referring to Q. 1, change the feedback resistor value to 22 kΩ. State the voltage, either side of zero, now required to make the output change state.

3 Draw the circuit diagram of an inverting Schmitt trigger based on the diagram in Figure 6.2. Change the feedback resistor (the upper one) to 100 kΩ. Find out the voltage needed at the input, either side of zero, that makes the output change state.

4 Change the 100 kΩ resistor in Q. 3 to 47 kΩ. What effect does this have on the behaviour of the circuit?

5 Draw the circuit in Figure 6.3 and add two buzzers, set at two different frequencies, to create a two-tone siren (no other components are necessary).

6 Design a voltage regulator, based on an op-amp, transistor and any other necessary components, that provides a 9 V output.

7 Explain why a 'balanced microphone' is likely to provide a signal less prone to picking up unwanted interference than an unbalanced microphone. Use a diagram to explain your answer. Why are the inner wires of a balanced cable twisted together?

Chapter 7

Logic gates

Which family to use?

Electronics Workbench allows us to use a wide range of logic gates (e.g. AND, OR, NAND, NOR, etc.). The two main families are TTL (transistor-transistor-logic) and CMOS (complementary metal oxide semiconductor).

We will avoid confusion at this stage by using the default gates, which are *ideal* – this means that their outputs can supply an unlimited current and their inputs require no current at all.

A real circuit requires a power supply. A 5 V supply will operate with all the logic gates available, and we will change the Workbench battery accordingly.

Taking shortcuts

We will begin with a circuit designed to show the action of an AND logic gate. Figure 7.1(a) shows the type of arrangement that could be used in practice. However, no logic gate could normally supply a bulb directly (remember, the Workbench gate is ideal) and a transistor output circuit would normally be required. When checking logic circuits we often wish to check just the action of the gates. Workbench allows us to take a shortcut by using a 'voltage source' and 'probe'. The voltage source is like a battery, except that it has only one connector and, similarly, the probe has a single input, but lights like a bulb.

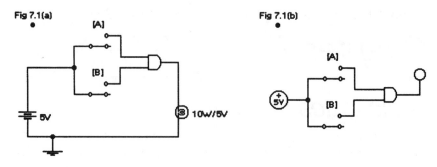

Figure 7.1

Try the circuit shown in Figure 7.1(b). Its action is identical to the battery-driven circuit, but no ground connection is required. Clearly, a real circuit could not be built this way, but complex arrangements of gates can be simulated faster, the diagram is quicker to make, and appears less complex.

Logic conventions

Logic systems work at two voltage levels hence 'binary logic'. These levels are referred to as logic 0 and logic 1. Logic 0 is represented by 0 V; logic 1 is represented by the positive supply voltage – in our case, 5 V. As there are only two possibilities, it does not matter if the voltages are not accurate – so anything below about 2 V counts as logic 0, and any voltage above about 3 V counts as logic 1.

The diagrams shown in Figure 7.1 assume that any input not connected to anything will be at logic 0. This is a dangerous assumption and, in real circuits, more care must be taken to fix the inputs to the required logic level. We have also assumed that, if the output is at (or near) 0 V, it represents logic 0. So, when the probe is *not* lit, it represents logic 0.

Truth table

A truth table sums up the behaviour of a gate. All the possible combinations of inputs are listed. For example, a gate with two inputs labelled A and B could be expressed as follows:

A	B
0	0
0	1
1	0
1	1

A zero in the truth table means 'switch open (off)' and a one means 'switch closed (on)'. If the lamp is unlit, it indicates a zero and, if lit, a one. The output logic levels (denoted by the letter Q) are:

A	B	Q
0	0	0
0	1	0
1	0	0
1	1	1

In other words, when switches A and B are both open (off) the output is at logic 0. When both switches are closed together, the output switches to logic 1. This is summed up by the expression: $Q = A \text{ AND } B$.

The logic converter

One of the most stunningly effective devices in Workbench is the logic converter. This is the instrument at the top of the screen to the right of the other instruments. It is shown in use in Figure 7.2

Figure 7.2

1 Drag the logic converter into position.
2 Connect up an AND gate as shown.
3 Double-click the converter and click the first CONVERSION, i.e. GATES to truth table. The truth table will magically appear.

Boolean algebra

Clicking the next conversion will turn the truth table into a mathematical expression known as Boolean algebra. A description of Boolean algebra is beyond the scope of this book, but it is basically a statement of the gate or system of gates expressed in the form 'Output will be logic 1 if ...' For example, the operation of an AND gate could be summed up as follows:

Output is logic 1 if input A AND input B is logic 1

that is:

$$Q = A \text{ AND } B$$

Note: in Workbench the 'Q=' part is omitted.

In Boolean algebra, a 'times' sign stands for AND, and a 'plus' sign stands for OR. This may seem odd at first, but it allows the use of all the normal rules of algebra when simplifying or expanding expressions. In practice, a dot usually represents the times sign, and in Workbench the dot is omitted; hence AB means A times B. So the AND gate becomes simply: AB. An OR gate is expressed as: A + B.

When a truth table is converted into a Boolean expression it is often possible to simplify the expression and thereby create a smaller circuit. This only applies to large systems of gates, of course. Clicking the next box causes the program to attempt to reduce the number of terms in the expression. This can save many hours of manual labour.

Boolean to truth table

It is possible to type in an equation and make Workbench convert it into a truth table.

1 Use the cursor to clear any letters already in the box at the bottom of the logic converter, and type in the expression: A + BC + AC. This means: Output logic 1 if A OR (B AND C) OR (A AND C) are logic 1.
2 Now click on the fourth conversion box from the top. The truth table should appear.

3 Now click on the next box down. A circuit will appear showing two AND gates and two OR gates. A careful check of the circuit will show that it does obey the expression. However, it is not the simplest circuit possible.
4 Try clicking on the third box down. The expression will now become BC + A.
5 Now click again on the fifth box down. A simpler circuit appears that consists of just one AND gate and one OR gate. It does exactly the same job, but with less circuitry.

In practice, gates are obtained in integrated circuits. For example, AND gates are available in an IC that houses four 2-input gates in one package. So our circuit would require two ICs. It would be helpful if we could change the circuit to use only one type of gate.

It is possible to build any logic circuit using only NAND or NOR gates. The action of these gates will be discussed later, but try clicking on the lowest box. A circuit will appear that does the same job as before, but consists only of NAND gates. The result consists of two NAND gates and one NOT gate (or inverter), which is made by joining the inputs of a NAND gate together. Three NAND gates are required and our circuit can be built using only one IC.

A coffee machine

A coffee machine must allow liquid to be poured if a coin has been inserted, or a token (a different slot), and if a button has been pressed, and if a cup is in place. We will assume that sensors are fitted to detect these conditions. The output will cause a solenoid valve to switch on the flow of liquid.

We can state the problem as follows:

Valve on if: (coin OR token) AND cup AND button

or:

(coin AND cup AND button)
OR
(token AND cup AND button)

Remembering that OR is written + and AND is a dot or no sign), we could have written:

$$Q = (\text{coin} + \text{token}) \text{ cup button}$$

or:

$$Q = (\text{coin cup button}) + (\text{token cup button})$$

Now we will insert letters to represent the sensors. The coin sensor will be A, the token B, the cup C and the button D, so:

$$Q = (A + B)CD$$

1 Clear the screen and try typing in the box at the bottom of the converter, the expression: (A+B)CD.
2 Now click the appropriate box to make the truth table appear.
3 Next click the box that converts the expression into a circuit. Notice how two 2-input AND gates are shown, although a single 3-input AND gate could be employed.
4 Click the box that converts the truth table to a simplified expression. The expression should become: ACD + BCD. The word 'simplified' is misleading, as it only means that the brackets are removed and the expression multiplied out. If converted to a circuit, it will appear more complicated! Try it and see. Again, the two pairs of 2-input AND gates could be replaced with 3-input AND gates.

How the output works

A common misconception surrounding logic circuits is that the output can be 'on' or 'off'. We will now see that the words on/off can be very misleading and should always be avoided.

Try the circuits shown in Figure 7.3. Notice how the inputs are tied together so that a single switch can be used to control the gates. It is important that bulbs and batteries are used in order to illustrate the point in question. The batteries and bulbs should be set to 5 V; it is wise also to reduce the power rating of the bulbs.

When the output of the gate is at logic 1 the bulb in the first circuit lights. However, when the output of the gate is at logic 0 the bulb in the second circuit lights.

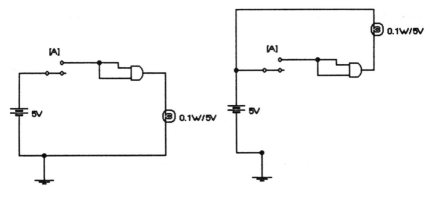

Figure 7.3

It is important to remember that the gates are connected to the power supply rails, even though the connections are often omitted in circuit diagrams. Hence, the output can supply (source) current when it is at logic 1, and can take in (sink) current when at logic 0.

Figure 7.4 illustrates the operation of the logic gate output in terms of a pair of transistors. Note that the switch in each circuit has been rotated so that the input can be joined to positive or 0 V. When the input is positive (logic 1) the upper transistor turns on and current flows through the bulb in the left-hand circuit. When the input is at 0 V the lower transistor turns on and current flows from the positive rail, through the bulb in the right-hand circuit.

A real logic gate employs a similar arrangement of transistors and, at any one time, either one transistor or the other is switched 'on'. Hence, the output is *always* 'on' in some sense, whether it is at logic 0 or logic 1. Figure 7.5 further illustrates the output used as

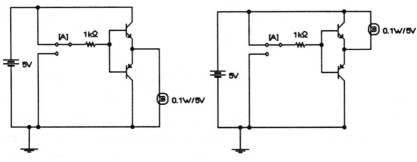

Figure 7.4

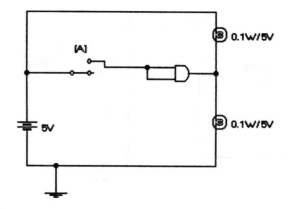

Figure 7.5

a source or sink. As the output is either positive or at 0 V, only one lamp will light.

Further gates

The logic gates available in Workbench are: AND, OR, NOT (sometimes called inverter), NAND, NOR, EXCLUSIVE OR, EXCLUSIVE NOR, Tristate buffer, buffer. The operation of all these gates can be explored by means of switches and indicator probes, or by employing the logic converter as described. In brief summary:

Gate	*Expression/Description*
AND:	Q=1 if A=1 AND B=1
OR:	Q=1 if A=1 OR B=1 OR both = 1
NOT:	Q=1 when A=0 ; Q=0 when A=1

This can be expressed by:

$$Q = \bar{A}$$

Where $\bar{A}$ reads as 'not A'. Note that Workbench writes this as A'.

NAND: The opposite result of using an AND gate. A NAND gate can be made with an AND gate, followed by a NOT gate as shown in Figure 7.6

AND NOT = NAND

Figure 7.6

NOR: The opposite result of using an OR gate. A NOR gate can be made with an OR gate, followed by a NOT gate as shown in Figure 7.7

OR NOT = NOR

Figure 7.7

Exclusive OR (XOR): Q=1 if A=1 OR B=1, but if *both* A and B=1, Q=0

Exclusive NOR (XNOR): Q=0 if A=1 OR B=1, but if both A and B=1, Q=1

Buffer: Q=1 if A=1; Q=0 if A=0

In other words the output copies the input. The buffer may not appear particularly useful, but the output current available may be many times more than the current required by the input.

Tristate buffer: acts like a normal buffer if the 'enable' input is at logic 1. If the enable input is at logic 0, the output is 'open circuit' meaning disconnected. This can be useful if several outputs are joined to a common connection (for example, a data bus in a computer). In a normal two-state system, if one output was at logic 1 and another output at logic 0, there would be a short circuit between the two outputs. Although logic gates will often tolerate this treatment, it will cause confusion in a circuit.

A tristate output allows the output to be at logic 1, logic 0, or disconnected. Workbench includes a tristate buffer, but in practice other tristate gates are available, such as a tristate NOT. Of course, any gate in Workbench can be made into a tristate version by connecting the tristate buffer in series with its output.

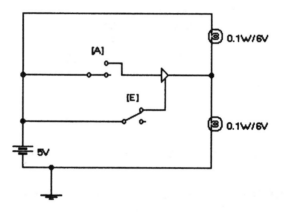

Figure 7.8

Try the circuit in Figure 7.8. Notice that the bulbs have been changed to 6 V types. This ensures that they are both unlit when the tristate buffer is disabled. Note that the output copies the input when the enable is at logic 1, but neither bulb lights when the enable input is at logic 0. In real life, the bulbs would be half lit (two bulbs in series) when the tristate output is disconnected.

Combining gates

We have already combined AND gates with OR gates in the coffee machine example. However, with the full range of gates, more ambitious circuits can be attempted.

Example 1

We want an automatic light to switch on when it is dark, but *not* after midnight. As before, we will call the output Q, so if Q=1 the light will be lit. Our inputs will be A and B. Hence A=1 when it is dark, and B=1 after midnight. We can write the equation:

$$Q = A \text{ AND } B'$$

The equation suggests that an AND gate and a NOT gate will be needed. If in doubt about the required circuit, try typing the equation into the logic converter. Remember that 'Q=' is assumed, so should not be typed, and that AND is omitted in Workbench – simply type AB'. Convert the equation to a truth table. Notice how the output is 1 only when A=1 and B=0.

Convert the equation to a circuit. The circuit shown in Figure 7.9 should appear. The NOT gate causes 'not B' to be applied to the AND gate. If the circuit was built, it would require two ICs (i.e. two chips) one containing NOT gates; the other AND gates. The cost would be halved if one IC was employed. Try clicking on the box that converts the equation into NAND gates: this shows that two NAND gates can perform the operation of an AND gate. The second NAND gate has its inputs joined together: this makes its operation identical to that of a NOT gate. The effect of inverting the output of a NAND gate makes it behave as an AND gate.

Figure 7.9

Workbench still retains the NOT gate just below input B. However, this can also be made from another NAND gate with its inputs joined together. The final solution is shown in Figure 7.10.

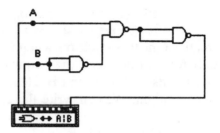

Figure 7.10

Example 2

We wish to water a garden automatically if it is dark OR if the override switch is set AND the soil is NOT wet. We will assume that the automatic water valve is open when Q=1.

- When it is dark, A=1
- When the override switch is set B=1
- When the soil is wet C=1

So our equation becomes:

$$Q = (A \text{ OR } B) \text{ AND } C'$$

That is:

$$Q = (A+B)C'$$

Convert the equation into a truth table by clicking the appropriate box. Next convert into a circuit. The circuit shown in Figure 7.11 should appear. It requires three different types of gate and hence three ICs. Try converting the equation to NAND gates. Three NAND gates and three NOT gates (i.e. NANDs with their inputs

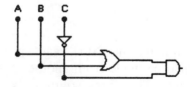

Figure 7.11

joined together) are required. Since there are four 2-input NAND gates contained in one IC package, two chips will be needed.

Try clicking on the box that simplifies the equation. Notice how the brackets are removed so that the equation reads:

$$AC'+BC'$$

Now convert this to a circuit. The result requires one more AND gate. The word 'simplify' is used loosely in this context! However, try converting the equation into NAND gates. This time, a much more useful circuit is presented, showing that the system can be constructed from just four NAND gates (remember that the NOT gate is made by joining the inputs of a NAND gate together).

Schmitt triggers

We have already met the Schmitt trigger in Chapter 6, but any non-inverting logic gate or system can be made into a Schmitt trigger. As with the op-amp, we need to apply positive feedback and a suitable circuit is shown in Figure 7.12. The logic gate must be a CMOS type (either 74HC or CMOS 4000) and it must be a non-inverting type, such as a buffer. It is possible to use an AND or OR if their inputs are connected together. Figure 7.12 also shows how two NOT gates can be made into a Schmitt. Again, the NOT gates could be made using NANDs or NORs with their inputs joined.

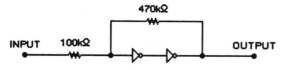

Figure 7.12

It is also possible to purchase ready-made Schmitts, in which case the resistors shown in Figure 7.12 are not required. Using a ready-made Schmitt is just like using a normal gate.

A Schmitt gate or system is ideal when a very clearly defined output is required. The output moves clearly between logic 0 and logic 1. Assuming a 5 V supply, the input has to move some way past 2.5 V before the output changes state.

A Schmitt system has many applications and one use of the logic gate version is to clean up digital signals that contain interference (noise). Stray signals are often picked up in a wire, e.g. mains cables produce magnetic and electric fields that induce signals into nearby wires. Generally, these unwanted signals will be quite small in voltage; providing they are less than about 3 V our Schmitt trigger will completely ignore them. Hence, the output from the Schmitt will contain only the digital information – the noise will have been eliminated.

Another application of the Schmitt trigger is to convert slowly rising or falling logic signals into fast-changing signals. The counters discussed in Chapter 11 require a clock signal that changes abruptly between logic 0 and logic 1. If this transition is slow or hesitant, the counter will produce erratic results. A Schmitt trigger neatly cleans up a digital signal where there is doubt regarding its quality.

The action of the Schmitt trigger can be demonstrated with a potentiometer and voltmeters as shown in Chapter 6. However, a more automated approach is shown in Figure 7.13. Here a signal generator (set to 'sine wave') is used to produce a slowly rising and falling signal. Change the amplitude of the signal to 2.5 V and the offset to 2.5. The oscilloscope shows the sine wave input compared with the square wave produced by the Schmitt trigger. Double-click the leads to the oscilloscope to change their colours for best results.

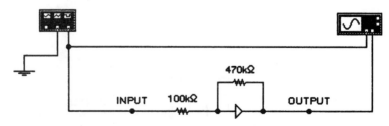

Figure 7.13

Questions

Full written answers, complete with Workbench circuits, are available on the accompanying disk. See p. 224 for details.

The questions in brackets are provided for you to answer in terms of Boolean algebra if you so wish.

1 Draw the truth tables for the following 2-input gates:

(a) AND
(b) OR
(c) NAND
(d) NOR
(e) XOR
(f) [Write the Boolean expressions for the gates.]

2 Name and draw the single gate that can replace an AND gate followed by a NOT (inverter) gate. [Show how Boolean algebra can produce the required answer.]

3 Name and draw the single gate that can replace a NOR gate followed by a NOT gate. [Show how Boolean algebra can produce the required answer.]

4 Draw a truth table to show the behaviour of the system in Figure 7.14. Draw a single gate that can replace the system shown. [Write the Boolean equations of both systems.]

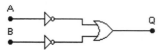

Figure 7.14

5 Design a logic system to sound a warning buzzer if the conditions in your tropical fish tank are incorrect. Various sensors are used as follows:

● Temperature too low; sensor A switches to logic 0
● Temperature too high, sensor B switches to logic 1
● Filter not working; sensor C switches to logic 1
● Light not working; sensor D switches to logic 0.

Begin by listing the conditions in the form of an equation. [Express this as a Boolean equation.]

Now draw the system of gates required to solve the problem. Assume that the final gate is capable of driving your buzzer directly. You may double-click a gate and change its number of inputs if necessary.

6 Cash-dispensing machines may check your identity by means of a thumb print or by shining a ray of light into your eye. Design a logic system that allows you to draw cash if BOTH conditions A and B are met, together with condition C or D:

- Card inserted; sensor A switches to logic 1
- Correct entry code; sensor B switches to logic 1
- Correct thumb print; sensor C switches to logic 1
- Correct eye match; sensor D switches to logic 1.

Begin by listing the conditions in the form of an equation. [Express this as a Boolean equation.]
Now draw the system of gates required to solve the problem.

7 Show how a 4-input OR gate can be made using 2-input OR gates. [Write the Boolean equation that describes the system.]

8 Show how a Schmitt trigger circuit could be made using only 2-input NAND gates. You need not enter the values of the resistors used.

9 Draw the system of gates required to satisfy the equation (use 3-input gates where necessary):

$$Q = A.B.C + A.C + B.C$$

(a) Draw the truth table of your system.
(b) Convert the system to one using only 2-input gates.
(c) Simplify the system by trial and error, or using Boolean algebra.

Chapter 8

Real logic circuits

Logic gates are sold in the form of ICs. These appear in a variety of forms, but the easiest to use are the DIL (dual in-line) packages, with pins down each side as shown in Figure 8.1. The pins are designed to fit on a 0.1 inch grid and can be directly inserted into a piece of stripboard (often known as veroboard), or a custom-designed printed circuit board (PCB).

Figure 8.1

All 2-input AND, OR, NAND and NOR gates are supplied in 14-pin DIL packages. As each gate requires 2 inputs and 1 output, four gates use a total of 12 pins, leaving 2 pins for the power supply connections.

Families

It is generally unwise to mix ICs from different families and therefore it is important to select the family required at an early stage in the design. There used to be a clear distinction between two family groups known as TTL (transistor-transistor-logic) and CMOS (complementary metal oxide semiconductor). TTL is based on bipolar transistors similar to types such as BC108. CMOS is based on transistors more akin to 'field effect' types. You will not see any of the transistors in either family group – they are buried within the silicon chip.

TTL gates used to have higher operating speeds when compared with CMOS. However, CMOS offered much lower operating power and tolerated a wider supply voltage range. The present situation has been complicated by the introduction of 'high speed CMOS' and 'low power TTL'.

TTL

A widely used TTL IC is known as the 74LS series. For example, a 2-input NAND IC is known as 74LS00. A 2-input NOR IC is 74LS02. This family requires a 5 V supply. The voltage accuracy required is important enough to require the use of a regulator. This, and the fact that a significant current is required, makes the IC generally unsuitable for battery powered projects.

An important point to note is that the inputs 'float high'. In other words if the inputs to a gate are not connected to anything, they are assumed to be at logic 1. A significant current (i.e. about 0.4 mA) is required to pull an input to logic 0. This compares unfavourably with CMOS, which requires virtually no current at an input. The output can generally source a current of less than 1 mA, though it can sink around 8 mA. Again, the source current is unfavourable compared with CMOS.

Testing a TTL LS gate

Try the circuit shown in Figure 8.2. Notice that it employs an AND gate, so both inputs must be at logic 1 for the output to switch to logic 1. The state of the output is indicated by the bulb. It must be set to 5 V, 0.1 W. We could not expect a real gate to supply enough current to light a normal bulb, but Workbench allows us to use ultra-efficient bulbs!

Select the gate by clicking once, then select MODEL from the circuit menu. This allows the gate – which at present is ideal – to become a member of a particular family. Now select TTL, then LS.

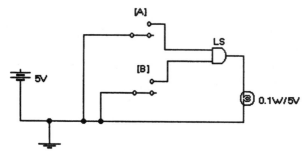

Figure 8.2

Activate the circuit. The bulb will light, even though both inputs are disconnected from the supply. Note that the battery positive appears to be disconnected from the circuit. In practice, it would be connected to the power supply connection of the IC.

It is essential to remember, with all gates, that they are connected to the power supply rails, and that any current flowing from the output is derived from the positive supply rail.

The circuit shows that, if the switches are open, the inputs to the gate are treated as logic 1 – even without connecting them to positive. To make the inputs logic 0, the switches must be closed.

Summary

We have seen that the TTL LS series is not well suited to battery powered circuits and the inputs are treated as logic 1 unless tied to logic 0.

CMOS ICs

The traditional CMOS family is known as the CMOS 4000 series and, despite being slower than its rivals, it still offers the best option for most battery powered project work. The CMOS 4000 series can operate on supplies of between 4 and 16 V. CMOS ICs consume very little power – in fact, a static CMOS circuit can remain powered by a small battery for up to a year.

CMOS inputs must be tied to a specific logic level. They must never be left 'floating', i.e. unconnected – even for a short time. Very little current is required by an input – so little that the static charge on your fingers can easily destroy a CMOS device. The tiny current requirement is generally an advantage, but CMOS ICs must be handled with care.

CMOS outputs can only reliably sink or source just over 1 mA. However, on a 12 V supply they can just light up an LED providing the output is not connected to other inputs.

A 2-input CMOS NAND IC is known as type 4011B. A 2-input NOR is type 4001B.

Best of both worlds

In an attempt to produce gates that combine the advantages of TTL with CMOS, i.e. high speed with low power consumption, a number of new families have been created. The most useful is the 74HC series, operating on a supply of between 2 V and 6 V. Their code numbers generally conform to the TTL system. Hence, 2-input NAND ICs are known as type 74HC00, 2-input NOR ICs are type 74HC02 and 2-input AND ICs are type 74HC08.

The inputs of the 74HC series have all the advantages of ordinary CMOS, though precautions are still required with regard to damage from static electricity. The outputs are better than any other series, and are capable of sourcing or sinking at least 5 mA – more if they are not connected to the inputs of other gates.

Summary

It is generally recommended that new designs are based around the 74HC series; however, the power supply range of 2 V to 6 V is less convenient than with CMOS 4000. If CMOS is selected in Workbench, the HC type will be assumed. Workbench also assumes a power supply of 5 V.

Testing a CMOS HC gate

Figure 8.3 shows how a CMOS IC can be connected. Note the use of 'pull down' resistors. These ensure that the inputs are at logic 0 unless the switches are closed. Workbench will excuse you if you forget the resistors, but they are essential in real life. The value of the resistors can be anywhere between 1 kΩ and 1 MΩ. If either input is connected to a clearly defined logic level, such as an output from another gate, then its 'pull down' resistor may be omitted.

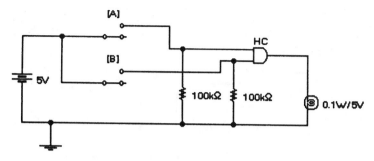

Figure 8.3

Do not forget to change the gate into a CMOS type by first clicking on the gate then selecting MODEL from the circuits menu. Select CMOS HC. Make sure that the lamp has a 5 V, 0.1 W rating and that the switches are connected to the positive side of the battery.

Making a real circuit

Suppose we wish to light a real lamp, which requires a much larger current than our logic gate can supply? The answer is to use a transistor as shown in Figure 8.4. In practice, a high-gain low-power transistor, such as BC108 or BC184, can be employed. The power of the lamp can now be raised to a more reasonable 1 W.

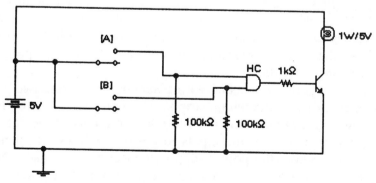

Figure 8.4

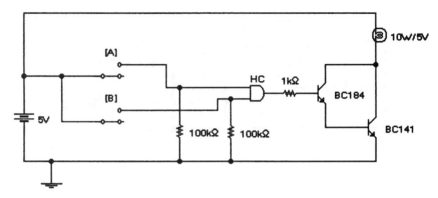

Figure 8.5

This is still rather low, and so Figure 8.5 employs a darlington pair. The first transistor is still the high-gain low-power type mentioned earlier, but the second is a higher power transistor, such as BC141. High-power transistors have a low gain, but this is of no consequence, as the gain will be equal to the gains of each transistor multiplied together. The rating of the bulb can now be raised to 10 W. Note that, in practice, transistor type TIP41A (not available in Workbench) would provide a higher power rating at little extra cost. As discussed in Chapter 3, a single darlington transistor, such as TIP121, can replace the two transistors shown.

Testing a real IC

Begin by choosing the IC family. Let's assume that a member of the 74HC series will be tested, such as 74HC08 (a quad 2-input AND gate). The data for an IC can be obtained from any good-quality electronics catalogue. The layout of this IC is shown in Figure 8.6. Note that the code 7408 refers to the layout of the IC and applies to 74LS08 *and* 74HC08.

Figure 8.7 shows the actual wiring diagram for testing two gates if the circuit is constructed on stripboard. Figure 8.7(a) shows how pull down resistors are employed to hold the inputs of the gates at 0 V. Connecting any of the inputs to positive will switch them to logic 1. Note how the power supply connections are made. Note also that all the input pins of the unused gates must be connected to logic 0 or logic 1; the inputs of a CMOS IC must never remain

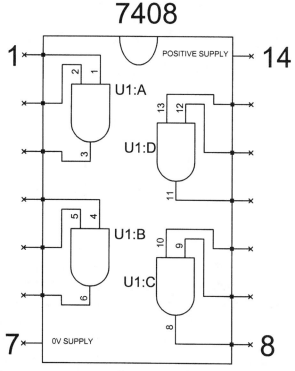

Figure 8.6

unconnected, even if a particular gate is not used. Hence, the inputs of the two gates on the right-hand side of the IC are all connected directly to 0 V. The unused outputs must remain unconnected.

Figure 8.7(b) shows the stripboard layout for testing the two gates on the left-hand side of the IC. Note that the stripboard tracks that run underneath the IC and hence join pin 1 to pin 14, pin 2 to 13, etc. must all be broken. The resistors that connect each input to 0 V may be of any value between $10 \text{ k}\Omega$ and $1 \text{ M}\Omega$. The resistors in series with the LEDs should have values around $330 \,\Omega$.

Anti-static precautions

The IC is supplied in a special anti-static tube or foam. Leave it in the original packing until required. Assuming that the circuit is soldered together, it is wise to use an IC socket. The IC can then be inserted when all soldering is complete.

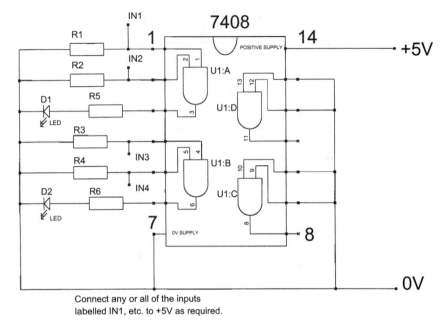

Connect any or all of the inputs
labelled IN1, etc. to +5V as required.

Figure 8.7a

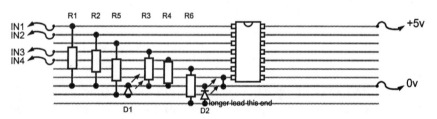

Figure 8.7b

Providing the circuit is constructed correctly and none of the IC inputs are unconnected, the only danger presented to the IC is during transfer from its protective packaging to the IC socket. Begin by touching an earthed metal object, such as the metal case of an appliance plugged into an earthed mains supply – this will remove any static charge in your body. Then remove the IC from its packaging and insert it the correct way round into its socket. Remember that pin 1 of the IC is to the left of the notch. Expensive ICs can be further protected by the wearing of earthing straps, which keep your body permanently earthed. Consult a catalogue for details.

Once the IC is safely installed, the components and connections in the circuit will prevent damage from static electricity, and the circuit may be handled without taking further precautions.

Example 1

This circuit is simply a test circuit to illustrate the action of a gate. Other ICs may be inserted into the socket to test different gates. We will now see how the example circuits in Chapter 7 can become working projects.

Referring back to Figure 7.10, our final circuit employed NAND gates, so reducing the cost, as only one type of IC is required.

Figure 8.8 shows how the original skeleton circuit is turned into a full working circuit. The two potentiometers labelled R and L represent the 'daylight control' and the LDR. The daylight control is a variable resistor that sets the point at which the circuit responds to the falling daylight. Pressing R or SHIFT R changes the setting in steps of 5 %. Pressing L or SHIFT L simulates the changing daylight level. As the level of daylight falls, the resistance of the LDR rises; this causes the voltage at input A to rise until it is 'seen' by the NAND gate as logic 1.

It is essential to change the gates into CMOS HC types. Check also that the bulb is set to 5 V. The switch labelled M is the 'midnight switch'. When open, the circuit responds correctly to the level of daylight. When the 'midnight switch' is closed, the bulb is

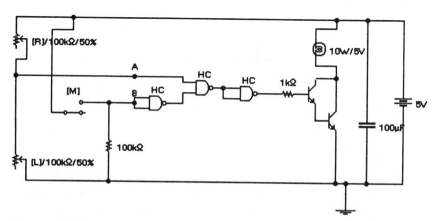

Figure 8.8

prevented from lighting. Notice the inclusion of the 100 kΩ resistor. This ensures that input B is never 'floating'. If the switch is open, the resistor pulls the voltage at input B to zero. The bulb is driven via a darlington pair, so that sufficient current is available. In practice a single darlington transistor could be employed, such as TIP121.

Finally, a 100 μF capacitor is connected across the supply rails to help smooth out voltage fluctuations. In larger circuits, two capacitors are best employed for this purpose: a larger value of, say, 1000 μF in parallel with 0.1 μF. The larger value – which will be an electrolytic type and which must be connected the correct way round with respect to positive and negative – removes larger voltage fluctuations and the smaller capacitor removes lesser voltage spikes.

This process is known as 'decoupling' but is unnecessary with Workbench and may slow down simulation.

If the circuit fails to work, use voltmeters connected to the outputs of the gates to check each stage of the circuit. The negative side of each voltmeter should be connected to 0 V in the circuit. The same fault-finding procedure applies in real life.

Example 2

We saw the solution in Figure 7.11. When the expression is simplified and converted into NAND gates, the circuit becomes the one shown in Figure 8.9.

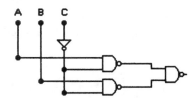

Figure 8.9

This may be converted into a working circuit in much the same way as with Example 1. The full circuit diagram is shown in Figure 8.10. There are many similarities with Figure 8.8 and so a full description is not be needed.

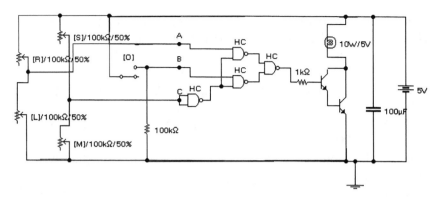

Figure 8.10

Potentiometer R simulates a daylight control setting, L simulates the LDR, M the moisture control setting and S the soil moisture sensor. In practice, this would be two wires pushed into the soil; moisture will cause the resistance between the wires to decrease. The switch labelled O is a manual override.

Begin with all variable resistors set to 50% and the override switch open. As it becomes dark, the resistance of the LDR (L) will increase. If the soil is dry, the resistance of S will increase. Try making L and S 55%. The lamp should light, showing that watering is taking place. As the soil becomes wet, the value of S will decrease. Try making it 45%; the bulb should switch off.

Change the value of S back to 55% (dry soil) and change the value of L to 45% to simulate increasing daylight. Watering should stop. Now close the override switch O; watering should begin again. The circuit fully satisfies the specification outlined in Chapter 7.

In real life, a solenoid control value would be substituted for the bulb. As this is likely to require a 12 V supply, it would be wise to power the whole circuit with 12 V, and use a CMOS 4011 IC instead of the 74HC00. A single darlington transistor could replace the two shown; TIP121 or TIP122 is suggested.

Questions

Full written answers, complete with Workbench circuits, are available on the accompanying disk. See p. 224 for details.

Assume a supply of 5 V in all questions, and note that every gate is connected to the positive and 0 V power supply rails. However, these connections are not shown.

1 (a) State an advantage of using a NAND gate TTL LS IC (e.g. 74LS00).

 (b) State two advantages of using a NAND gate CMOS IC (e.g. 74HC00)

 (c) If the inputs of your TTL gate are left unconnected, will the output be at logic 0 or logic 1?

 (d) If the inputs of a CMOS gate are left unconnected, what will happen?

2 CMOS ICs require special handling. State the precautions you would take when using a CMOS IC in your circuit.

3 A circuit may require an IC in a DIL package. Explain the meaning of DIL.

4 How is pin number 1 identified in a DIL layout? Draw an 8-pin DIL IC, labelling all the pins.

5 State the power supply voltage that must be used with:

 (a) TTL 74LS circuits

 (b) CMOS 74HC circuits

 (c) CMOS 4000 circuits.

6 Design a circuit that sounds a buzzer if switches A and B are both closed, but switch C is open. Use logic gates with a maximum of 2 inputs, a transistor, a 5 V battery and any other components to make a practical working circuit.

7 Design a combination circuit, using gates with no more than 2 inputs, which opens a solenoid lock (represented by a bulb) if switches A and B are closed, but if switches C or D are closed at the same time, the lock does not operate. Use a darlington pair output to drive a 10 W, 5 V bulb.

8 Add an alarm (i.e. a buzzer driven via a darlington pair) to your circuit in Q. 7, so that, if either switch C or D is closed, the alarm sounds. (Set the buzzer voltage to 4 V.)

Chapter 9

Logic gate multivibrators

A multivibrator circuit provides an output that changes abruptly between logic 0 and logic 1. There are three types of multivibrator:

1 Bistable: a circuit that is stable with its output at logic 0 or logic 1.
2 Monostable: a circuit that is stable only in one state. If it is forced into the other state, it returns to the stable state after a time delay.
3 Astable: a circuit with an output that does not remain in either state. It continuously changes between logic 0 and logic 1.

Bistable multivibrator

A simple form of latching circuit can be created with a single OR gate as shown in Figure 9.1. Note that the battery and bulb ratings are changed to 5 V, and the bulb set at 0.1 W. In this and all circuits in this chapter, the gates are CMOS HC types and resistors are used to pull floating inputs to 0 V. Workbench will forgive you if these resistors are omitted, but in real life they must always be fitted.

When the switch is closed, the bulb lights; when the switch is opened the bulb remains lit, i.e. the circuit is latched at logic 1. The latching action is caused by the use of *feedback*, i.e. the output is fed back to the spare input.

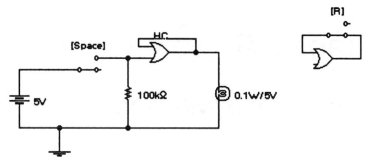

Figure 9.1

The inset diagram shows how a second switch can be added to the feedback loop to allow the circuit to be unlatched. Label the second switch R (for reset).

Note that the output from the logic gate is required to do two jobs: it must feedback a logic 1 to its spare input *and* light the bulb. If too much current is drawn from the output, its voltage will fall and this may affect its ability to remain latched. Such a system should never be used to power bulbs, or even LEDs, directly.

Figure 9.2 shows how a transistor may be added if an LED or bulb is required. The single transistor could be BC108 or BC184. If a current of more than 100 mA is required, then use a darlington transistor such as TIP121. Note the choice of 4.7 kΩ resistor in series with the transistor base. This value ensures that the current required from the output is not sufficient to affect its logic level.

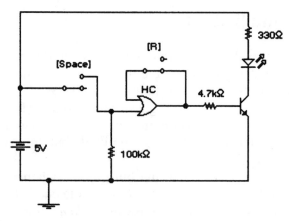

Figure 9.2

Other latches

A single buffer gate may be used as part of a latching circuit as shown in Figure 9.3. The diode is necessary to prevent current flowing from the switch directly to the output. A similar arrangement made with two NOT gates is shown in Figure 9.4.

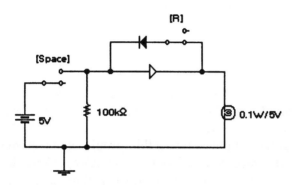

Figure 9.3

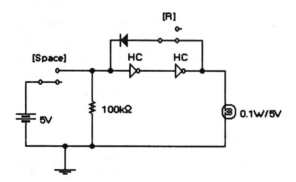

Figure 9.4

Remembering that NOT gates can be made using NOR gates, the circuit shown in Figure 9.5 provides an alternative latching circuit. It has the considerable advantage that a *positive* pulse applied via the reset switch causes the circuit to unlatch. Hence, the first switch acts as a 'set', and the second as a 'reset'.

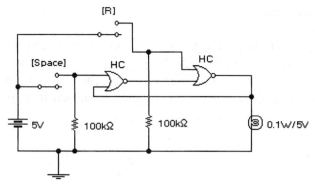

Figure 9.5

Professional bistable

The most professional version of this circuit requires two NOR gates or two NAND gates. We will look at the NOR gate version first. The circuit is sometimes referred to as a flip-flop. It has set and reset inputs and two outputs, which are always at opposite logic levels.

The circuit is shown in Figure 9.6. Try the circuit, but note that an odd phenomenon occurs when the circuit is first activated – the bulbs are likely to flicker on and off. Press R or S, noting that, in real life, momentary push switches would be used, so that R or S should be pressed twice to be left in their open states. The bulbs should stop flickering and the circuit will behave normally. The reason for the flickering is that both outputs try to attain logic 1 at the same moment, but as the feedback arrangement prevents this,

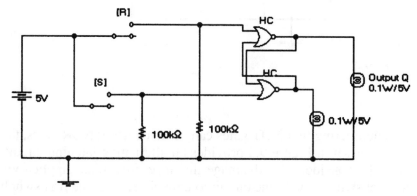

Figure 9.6

the circuit becomes unstable. The same problem can occur in real life if the two gates are identical, i.e. they operate at exactly the same speed. In practice this is unlikely, but it *can* occur and so provision should be made to force the circuit into a predictable state at switch on.

If a capacitor (say 0.1 µF) is connected in parallel with the reset switch, a pulse will be applied – at the moment of switch on – to the appropriate input, forcing the circuit into a known state. This does not work with Workbench, as it is not possible to control the power supply connected directly to the gates. (Remember that all the logic gates are connected directly to positive and 0 V, even though the connecting wires are not shown.)

Using Workbench, it is only necessary to toggle one or other switch to make the circuit behave normally. Note that closing the set switch for a moment makes output Q switch to logic 1; closing the reset switch for a moment causes the other lamp to light. This output is often called 'not Q', or $\bar{Q}$.

The switches labelled S and R are used to apply a positive pulse to the appropriate input. This pulse can be derived in a number of alternative ways, as described later.

Nand gate bistable

The NAND gate bistable or flip-flop circuit is shown in Figure 9.7. Note that the set and reset switches have been interchanged, the

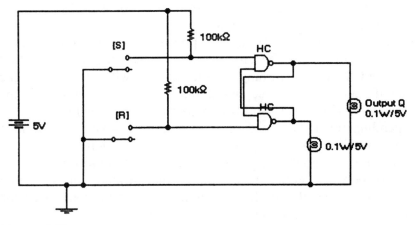

Figure 9.7

inputs are tied to positive via resistors, and each switch provides a negative pulse to change the state of the circuit.

Practical bistables

Any of the circuits described can be converted into practical circuits in the same way as described in Chapter 8. It is important not to draw too much current from any output and so a series resistor of value $4.7\,k\Omega$ and transistor should always be employed, unless of course the output is connected to the input of another logic gate. The addition of decoupling capacitors, also described in Chapter 8, will ensure reliable operation.

If long wires are used to connect the switches, there is a risk that interference will be picked up, and this may be sufficient to change the state of the circuit. This could easily happen with an alarm system where the wires might run a very long distance from the circuit to the alarm trigger switch. The solution is to connect a capacitor (say $0.1\,\mu F$) in parallel with each pull down or pull up resistor. The effect of the capacitor is to shunt away any alternating currents (i.e. noise) without affecting the change of DC that occurs when either switch is triggered. If the capacitor described earlier has been fitted to force the circuit into its reset condition at power up, this capacitor value should be increased to, say, $0.47\,\mu F$, otherwise its effect will be reduced by the 'noise cancelling' capacitor.

The full NOR circuit with these additions plus a buzzer output is shown in Figure 9.8. The transistor should be a BC184 in order to provide the gain required for the buzzer. Remember that the Workbench logic gates are always powered, even when the master switch is off, hence the circuit may still be unstable when first activated. However, turning on the master switch will allow the $0.47\,\mu F$ capacitor to cause an automatic reset. The addition of the capacitors will cause Workbench to run more slowly. In fact – depending upon the speed of your computer – you will be able to study what really happens during the first microsecond or so of turning on your circuit! A voltmeter (or oscilloscope) connected between either gate output and ground can be quite revealing, especially if the circuit fails to work. Remember the importance of the ground symbol, connected to the negative side of the battery.

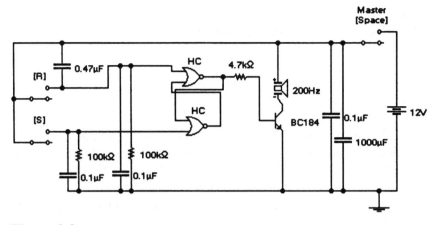

Figure 9.8

Monostable multivibrator

A monostable circuit is stable only in one state. The output can be
made to change state, but returns after a time interval to its original
state. A typical monostable circuit is shown in Figure 9.9.

Close the switch – the bulb should light. There may be a delay,
unless your computer is super-fast; this is simply a processing delay.
Once the bulb is lit, open the switch. The bulb should remain lit for
about a second (i.e. a real second – which may actually be longer on
the computer, depending upon its speed).

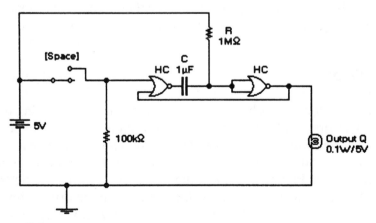

Figure 9.9

To sum up, closing then opening the switch will trigger the circuit into its unstable state, and the bulb will light. After a period of time (T), the circuit returns to its stable state. The time period T is given by:

$$T = 0.7 \times R \times C$$

where T is measured in seconds, R in ohms and C in farads.

R is invariably a very large number and C a very small number, but if the value of R is in MΩ and the value of C is in μF, the 'M' will cancel with the 'μ' making the calculation much easier.

So, with the values in Figure 9.9, the time period will be given by:

$$T = 0.7 \times 1\,M \times 1\,\mu = 0.7 \text{ seconds}$$

If the value of R was 330 kΩ and C was 47 μF:

$$T = 0.7 \times 0.33\,M \times 47\,\mu = 10.9 \text{ seconds}$$

Practical monostables

In practice, the value of the capacitor (particularly larger types) is quite inaccurate, so if an exact time period is required, R should be a variable resistor that can be 'fine tuned' to provide the time period required.

The precautions regarding decoupling and noise all apply to the monostable and, as with the bistable, the output should not be required to provide more than a very small current.

Egg timer

The egg timer circuit is based on two monostables as shown in Figure 9.10. The first is a timer for, say, three minutes (we will choose a much shorter time for testing) and the second monostable causes the buzzer to sound for five seconds. The values selected for the timing components in Figure 9.10 allow the whole cycle to be observed in a short time. For testing purposes, the buzzer may be driven directly from the output of a NOR gate; however, the buzzer voltage must be reduced to 4 V and its current rating to 0.0005 A.

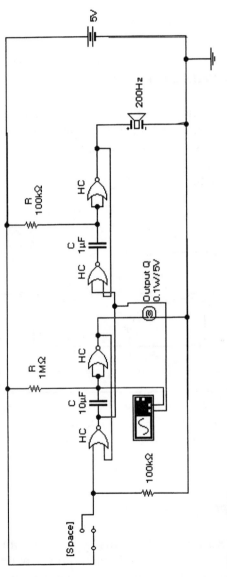

Figure 9.10

The switch provides a positive pulse for the first monostable. It is important for correct operation that the switch is closed, then opened. When the voltage at the output of the first NOR gate switches from 0 V to 5 V, the second monostable is triggered and the buzzer sounds.

Astable multivibrator

A typical astable is shown in Figure 9.11. Note that, although NOR gates have been employed, their inputs have been joined together to make them behave as NOT gates. We could have used NOT gates, or NAND gates with their inputs joined.

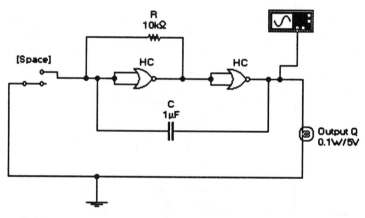

Figure 9.11

Getting jammed

A serious problem occurs with this type of circuit. If both gates operate at exactly the same speed, when power is first applied both outputs will try to switch to logic 1 at the same moment. The result is a jam! This always happens with Workbench, as the gates are modelled to perfection. In real life, jamming is less likely, though it *can* happen and the circuit should be designed to prevent the problem.

The solution with Workbench is to place a switch in a position where the output from one of the gates can be forced into a particular state. Activate the circuit with the switch open. Note how the bulb appears to be flashing normally. However, the system is unstable – the time counter will show the computer is labouring, the bulb will appear to flash at the same rate regardless of the values of R and C, and the oscilloscope will display an unhealthy and ragged wave.

Now close the switch to force the bulb into its unlit state. Open the switch. The bulb now flashes more slowly and a square wave will be displayed on the oscilloscope. The circuit is now working properly and the flashing rate can be controlled by the values of R and C. *Note:* Owners of fast computers should use a higher value of, say, 10 µF for 'C'.

Automatic self-starting

The use of the 'start switch' is cumbersome and a more intelligent arrangement is shown in Figure 9.12. In this circuit the positive supply rail is controlled via an on/off switch. Note that Workbench does not allow us to control the power supply to the logic gates – as soon as the circuit is activated the gates are powered up. So, when using Workbench, we have to ignore the odd behaviour of the circuit – which appears to operate even with the main switch open. Activate the circuit with the switch open, wait for the circuit to

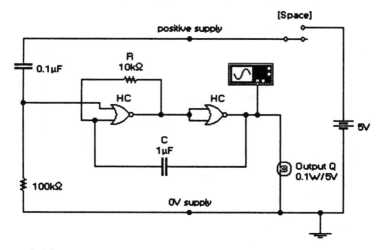

Figure 9.12

stabilize (as shown by the time counter counting up) then close the main switch. The circuit should now behave correctly as shown by the square wave display on the oscilloscope.

In real life, the gates would be powered from the points labelled 'positive supply' and '0 V supply', so the circuit will operate correctly.

The self-start components comprise the 0.1 µF capacitor and 100 kΩ resistor. If one of the inputs to a NOR gate is tied to 0 V, the other input can be used like the single input of a NOT gate. However, if an input to a NOR gate is tied to positive its output will be forced into a logic 0 state. Hence, the 100 kΩ resistor is used to maintain the spare input at 0 V, allowing the other input to form part of the astable circuit. But, at the moment of power-up, current will flow into the 0.1 µF capacitor, causing the voltage at the spare input to rise to positive. A moment later, the 100 kΩ resistor will cause this voltage to fall to zero, allowing normal operation – but the gate will have been 'kicked' into action. The value of 'C' should be increased to observe the bulb flashing on fast computers.

Perfect square wave generator

It is possible to design a circuit that produces a very good square wave and is always self starting – even on Workbench! However, a third gate is necessary. NOR gates are shown in Figure 9.13, but as their inputs are joined together they are actually NOT gates. NOT

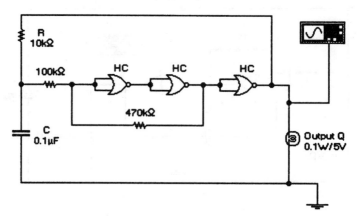

Figure 9.13

gates or NAND gates (with their pairs of inputs joined) will work equally well.

The first two gates, together with the $100\,\text{k}\Omega$ and $470\,\text{k}\Omega$ resistors, form a Schmitt trigger. It is the action of the Schmitt that makes the circuit self-start. The resistor labelled R, together with the capacitor labelled C, sets the frequency of the astable. The higher the value of R, and/or the higher the value of C, the lower the frequency will be.

Questions

Full written answers, complete with Workbench circuits, are available on the accompanying disk. See p. 224 for details.

1 (a) Draw an OR gate and show how it can be made to latch on when a switch (A) is closed, then opened. (Use a +5 V source and a probe to show the state of the output.)

   (b) Add a second switch (B) to your circuit that causes the gate to unlatch when the switch is closed, then opened.

2 Draw the 'professional' bistable circuit based on NOR gates. Your circuit should include a set switch, reset switch and 2 outputs labelled Q and Q'. (Use a +5V source and probes to show the output states.)

3 Draw the 'professional' bistable circuit based on NAND gates. Change the NAND gates to TTL LS, noting that, unlike CMOS gates, their inputs float high, i.e. they are at logic 1 if unconnected. You should therefore be able to design your circuit without any resistors, but you will need a ground symbol.

4 Design a timer (monostable) circuit based on NOR gates that sounds a buzzer (set to 4 V) driven via a transistor. Change the transistor model to BC184 to achieve a higher gain. Draw the full circuit with a 5 V battery and any necessary resistors. Calculate the time period obtained with a $100\,\text{k}\Omega$ timing resistor and $1\,\mu\text{F}$ timing capacitor.

5 Design an astable circuit based on NAND gates that is always self-starting. Use two BC184 transistors to drive two buzzers set to different frequencies to make a two-tone siren. Employ a 5 V battery and set the buzzers to 4 V.

Chapter 10

The 555 timer

The 555 timer, often labelled NE555, is a specialized IC designed to create monostable and astable circuits. It offers quite accurate time periods, though its accuracy still depends upon a timing capacitor and it cannot compete with crystal-controlled timing circuits. However, its time period is independent of the supply voltage – a great advantage over CMOS logic gate timers and clock generators in battery powered equipment. The general purpose NE555 requires a much larger operating current than a CMOS logic IC, though a CMOS 555 timer IC is available if required. The output from pin 3 of the 555 IC is very clearly defined – either logic 0 or logic 1. It never hesitates between these two levels.

The 555 timer offers a cheap and simple solution if a timer (monostable) or clock pulse signal (astable) is required. It suffers from a tendency to cause voltage fluctuations in the power supply rails, but remains very popular in educational establishments.

Facts and figures

- Supply voltage: 4.5 V to 16 V.
- PIN 1 (GND): power supply pin, 0 V.
- PIN 2 (TRI): trigger (starts timing cycle when it receives a 0 V pulse).
- PIN 3 (OUT): output (can sink or source up to 200 mA).
- PIN 4 (RES): reset (resets when it receives a 0 V pulse).

- PIN 5 (CON): control (offers fine control over the time period when connected to a fixed voltage. If not required it can be left unconnected. In sensitive circuits it can be connected via a 100 nF capacitor to 0 V for extra decoupling.)
- PIN 6 (THR): threshold (measures the voltage on the timing capacitor).
- PIN 7 (DIS): discharge (discharges the timing capacitor at the end of the time period).
- PIN 8 (VCC): power supply pin, positive.

555 monostable

The circuit in Figure 10.1 shows the basic method of connecting the 555 timer as a monostable or timer. Note that the pin layout is as it appears in real life, with pin 1 at the top left-hand side, and pin 8 at the top right. Try the circuit, noting that the bulb power is reduced to 1 W. To trigger the circuit, press the space bar twice to turn the switch on, then off. This provides a 0 V pulse at the trigger pin to start the timing cycle, but allows the trigger pin to return to positive.

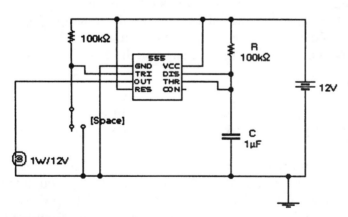

Figure 10.1

Time period

The time period during which the output pin switches to positive is given by the formula:

$$T = 1.1 \times R \times C$$

where T is the time in seconds, R is in ohms, and C in farads.

As before, the formula is equally valid if R is measured in $M\Omega$, providing C is measured in μF.

So, with the values given in Figure 10.1, $R = 100\,k\Omega = 0.1\,M\Omega$, and $C = 1\,\mu F$, hence $T = 1.1 \times 0.1 \times 1 = 0.11$ seconds (or 110 ms).

Try increasing the time period by increasing the value of R. If your computer is fast enough to work in real time, the value of R could be raised to $1\,M\Omega$ to provide a time period of just over one second. Try increasing the value of C.

How the circuit works

When power is applied to the IC, the output is at 0 V as is the discharge pin, so ensuring that the capacitor is fully discharged. When pin 2 receives a 0 V pulse, the output switches to positive for a time T. During this time, the discharge pin is allowed by the IC to float (i.e. it internally disconnects itself), so the capacitor C begins to charge via resistor R. The threshold pin monitors the voltage across the capacitor and, at two-thirds of the supply voltage the time period ceases. The output pin returns to 0 V, as does the discharge pin, so reducing the charge on the capacitor to zero, ready for the next cycle.

Sinking and sourcing

The circuit in Figure 10.1 employs the output as a source. In other words when the output is at 0 V no current flows through the bulb because both sides of the bulb are connected to 0 V. When the output switches to logic 1 (positive), current flows from the output and through the bulb to 0 V. This means that the bulb is lit during the time period.

Try the arrangement shown in Figure 10.2. This is the same circuit, but the output is now employed as a sink. When the output is at 0 V, the bulb lights, but when the output switches to positive the bulb is extinguished – the bulb lights before and after the time period.

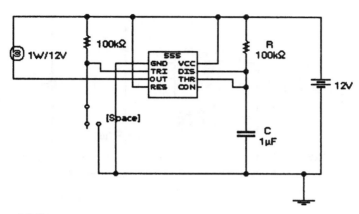

Figure 10.2

Auto start timer

The circuit shown in Figure 10.2 can make quite a useful timer, especially if the bulb is replaced with a buzzer. A main on/off switch is required to prevent the buzzer sounding continuously. If the existing switch is replaced by a capacitor it is possible to make the circuit start automatically whenever power is applied as shown in Figure 10.3. Try this circuit noting that the main switch must remain on throughout the timing period.

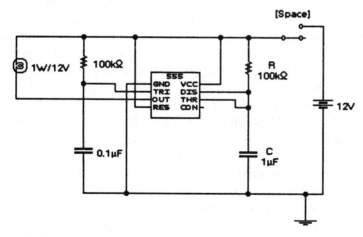

Figure 10.3

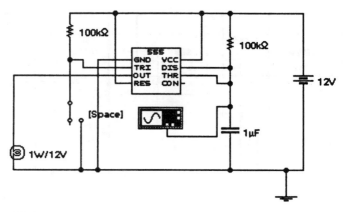

Figure 10.4

An oscilloscope may be connected, as shown in Figure 10.4, to monitor the charge on the capacitor.

Further work

The auto start timer shown in Figure 10.3 causes the buzzer or bulb to operate after a certain time, and remain on until the switch is opened. Design a timer, using two 555 timers, which cause the buzzer or bulb to operate for only a limited time, i.e. switch closed; main timer starts; after time period, bulb lights; after short time, bulb is extinguished.

The solution is provided in Figure 10.5. The main time period is provided by the values of R4 and C4. The output from IC2 is used to trigger IC1, which in turn operates the bulb or buzzer for a short time. Notice that the output from IC2 is not directly coupled to the trigger of IC1. This is because the trigger must remain positive for most of the time, apart from the brief 0 V pulse necessary to activate the IC. This brief pulse is provided via C2; for most of the time R1 keeps the trigger input of IC1 positive, but when the output of IC2 changes from positive to 0 V, this change of voltage passes via C2 to trigger IC1. A short time later, R1 returns the trigger input to positive again. This type of arrangement is known as 'AC coupling' and offers a reliable way of ensuring that the trigger input can return to positive after triggering. Note that the bulb lights briefly when the switch is closed, in addition to lighting at the end of the cycle. This can be viewed as an irritating fault, or a useful indicator that the circuit is

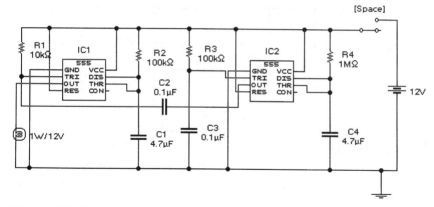

Figure 10.5

working! It is possible to purchase a single IC, known as the NE556, which contains two NE555 timers.

Other options

The circuit in Figure 10.6 shows how the reset control pins may be used. Instead of connecting the reset pin directly to positive, it is now tied via a resistor. This allows the use of switch X, which pulls the reset pin to 0 V to cause a reset.

To start the circuit, press the space-bar twice to apply a 0 V pulse to the trigger pin. Now press X twice to apply a 0 V pulse to the reset pin. The lamp should switch off immediately.

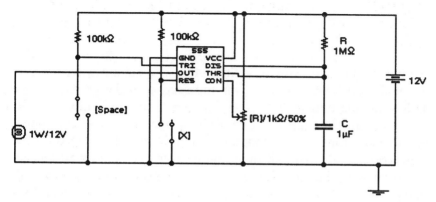

Figure 10.6

The diagram also shows how a potentiometer may be employed to set the voltage at the control pin. Try pressing R or SHIFT R to change the setting of the potentiometer. Note the effect that different settings have on the time period.

555 astable

The 555 timer IC can be wired as an astable, as shown in Figure 10.7. The resistors R1 and R2, along with C1, govern the overall frequency provided. The oscilloscope provides a reliable guide to

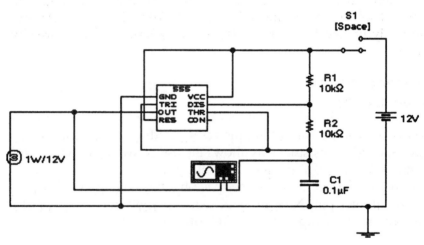

Figure 10.7

the behaviour of the circuit – the output is a square wave with an unequal mark/space ratio, i.e. the top section of the graph is twice as long as the bottom section – this means that the bulb is on for twice the time that it is off. The oscilloscope also shows the charge on the capacitor C1. Note that the switch S1 must be opened for the simulation to stabilize – if the program is activated with S1 closed, the bulb will flash randomly and the elapsed time display will remain blank. Fast computer users': increase C1 to 47 µF and set the oscilloscope timebase to 0.50 s/div.

How it works

When S1 is closed, the lack of charge on C1 will trigger the IC and make its output switch to positive. The bulb lights and the discharge pin is made inactive by the IC; current flows through R1 and R2 to charge the capacitor. When the voltage across C1 is two-thirds of the supply voltage, the IC changes state, making its output and discharge pin fall to 0 V. The bulb switches off and the capacitor discharges into the discharge pin via R2. The trigger pin is connected to the capacitor, so when the voltage across C1 falls to one-third of the supply, the IC is triggered again and the cycle repeats as before.

Notice that the capacitor charges via both R1 and R2, but discharges via only R2 – this explains the unequal mark/space ratio. If a more equal ratio is required then R2 should be much higher in value than R1; try raising R2 to 1 MΩ. After the first cycle (not normally observable in real life) the mark/space ratio is virtually equal.

Calculations

The fact that the mark/space ratio may be unequal complicates the calculation of the time periods and frequency, but the following formulae are provided for those not afraid of a little mathematics. Alternatively, try some values using Workbench, noting the results on the real time counter at the top of the screen.

$$\text{Mark time} = 0.7 \times C \times (R1 + R2)$$

$$\text{Space time} = 0.7 \times C \times R2$$

$$\text{Total period} = 0.7 \times C \times (R1 + 2R2)$$

$$\text{Frequency} = 1.4/[C \times (R1 + 2R2)]$$

If R2 is much larger than R1, then the period and frequency are largely dependent on R2. The formulae can be simplified as follows:

$$\text{Period} = 1.4 \times C \times R2$$

$$\text{Frequency} = 0.7/(C \times R2)$$

Alternative astable

It is possible to charge and discharge the capacitor from the output pin of the IC, as shown in Figure 10.8. The circuit requires less components and offers an equal mark/space ratio. However, it is unwise to use the output to trigger the IC unless it is employed only for driving a bulb, LED or transistor. Even then, it is important not to allow very much current to flow from the output. The time period is now simply given by:

$$T = 1.4 \times C1 \times R1$$

and the frequency is given by:

$$f = 0.7/(C1 \times R1)$$

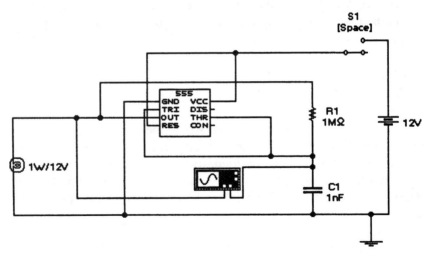

Figure 10.8

Design problem

Design a circuit using two 555 ICs that causes a lamp to flash at a frequency of 1 Hz for a total time of five seconds after a switch is turned on. There is more than one solution, but a possible circuit is provided in Figure 10.9. IC2 provides the monostable period of about five seconds. The output from IC2 is used to control the reset input of IC1. When the reset is positive, IC1 operates normally, but when its reset pin is held at 0 V it becomes jammed, with its output at 0 V.

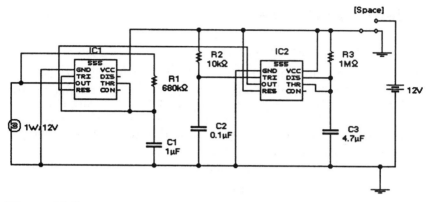

Figure 10.9

The circuit around IC2 includes the 'auto start at power on' components, namely R2 and C2. The values of R3 and C3 provide a period of about five seconds. IC1 is the astable arrangement shown in Figure 10.8, and the values of R1 and C1 provide a frequency of about 1 Hz.

If your computer cannot operate in real time it may be necessary to reduce the values of C1 and C3 by a factor of 10 or 100, to observe the circuit operating.

One final point: a ground symbol has been added to the spare pole of the switch to ensure that C2 is discharged when the circuit is switched off. In real life, this would not be necessary as the charge would leak away through the ICs, etc.

Questions

Full written answers, complete with Workbench circuits, are available on the accompanying disk. See p. 224 for details.

1 Show how a 555 timer can be used as a monostable (timer). Include a master on/off switch, another switch labelled 'start' and a 12 V battery. Arrange your circuit so that a bulb lights up when the start switch is closed and opened.

   (a) Calculate the time period if your timing resistor value is 470 kΩ and your timing capacitor is 10 μF.

(b) Calculate the resistor value necessary to achieve a time of 20 seconds if the timing capacitor value is 220 µF.

(c) Sketch a graph to show the capacitor charging.

2 Using the circuit from Q. 1 as a starting point, remove the bulb and connect a buzzer in such as way that the buzzer sounds at the beginning and end of the time period.

3 Design an astable circuit that makes a buzzer switch on and off repeatedly, but arrange the circuit so that the buzzer is on for much longer than it is off. State the mark/space ratio achieved.

4 Design a two-tone siren, based on a 555 astable, with a nearly equal mark/space ratio.

5 Refer to Figure 10.7. Assuming that R1=330 kΩ, R2=680 kΩ and C=0.01 µF, calculate:

(a) the time for which the output is positive (i.e. the mark time)

(b) the time for which the output is at 0 V (i.e. the space time)

(c) the total period of the output

(d) the frequency of the output.

Chapter 11

Flip-flops, counters and shift registers

A bistable or flip-flop circuit based on a pair of NOR gates was shown in Figure 9.6. The circuit had two switches labelled 'set' and 'reset' and two outputs labelled Q and Q' (not Q). In its reset state, output Q was at 0 V and Q' was positive. When the set switch was pressed for a moment, outputs Q and Q' both changed state. Pressing reset made the circuit revert to its original state.

The bistable system is at the heart of all binary counters, calculators, computers, etc. Not surprisingly, it is possible to buy the system as a ready-made IC.

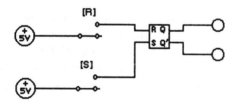

Figure 11.1

Look at Figure 11.1. The complete circuit has been reduced to a box with labels S, R, Q, Q'. To simplify the circuit, we have used 5 V voltage sources and logic probes. Try the circuit, noting that S or R must each be pressed twice to simulate the action of a momentary switch.

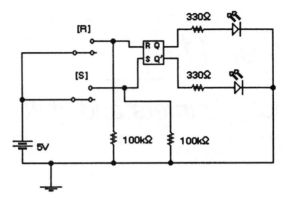

Figure 11.2

In real life, a complete circuit would be required and Figure 11.2 shows how a 5 V battery and LEDs with series resistors could be used. Note also that, if CMOS ICs are employed, the 100 kΩ pull down resistors are also essential.

From now on we will use voltage sources and probes, as these improve the clarity of diagrams, and pull down resistors will be omitted, as Workbench assumes that an unconnected input will be at logic 0.

D-type bistable

The D-type bistable employs some extra circuitry, which makes it far more useful than an ordinary bistable. Look at the circuit of Figure 11.3: there are two D-type bistables on offer in the parts bin; select the one with the 'bobble' at the top and bottom. The 'bobbles' mean that the connections operate with inverse

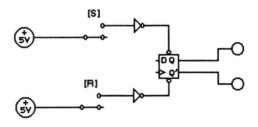

Figure 11.3

logic, so logic 0 causes a reaction, rather than logic 1. This complicates the situation, but the use of NOT gates, as shown in Figure 11.3, solves the immediate problem. The circle on the top of the D-type is the set input, and the circle at the bottom is the reset input.

Try the circuit, noting that it behaves exactly like the one in Figure 11.1. There are two extra connections which extend the range of possibilities enormously: one is labelled 'D' (data), and the other has a triangular symbol that means 'clock input', sometimes labelled CP (clock pulse) or CLK (clock). We have already made some clock pulse generators: the astable circuits in Chapters 9 and 10 produce square wave outputs that are ideal for driving clock inputs. A clock pulse is a regular signal – in an electronic watch, a crystal generates an accurate clock signal of 32 768 Hz (pulses per second). Clock pulses are used in calculators and computers, so that numbers can be processed in an organized way. At each clock pulse, a number might be added to memory, or the next step of a program might be executed. The faster the clock pulse, the better the machine in terms of number-crunching power. Generating a high-speed clock pulse is one thing; developing an IC that can operate at very high speeds without using too much power, or overheating is another. Hence, the step-by-step approach to the development of ever faster computers.

Operating the D-bistable

The D-bistable allows data to be moved from one place to another *in step* with the clock pulse. Figure 11.4 is a test circuit that shows the principle: begin by removing the D-bistable symbol with set and reset (these complicate the situation, unless they are both tied to 5 V) and use the D-bistable without set/reset connections.

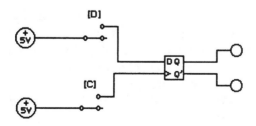

Figure 11.4

Note that pressing the switch labelled D (data) has no effect on the output. But if D is *left* closed the outputs will change state when C (clock) is closed. If C remains closed, D has no further effect. The situation is summed up as: output Q copies the logic level at D at the *moment* when C is closed. In other words, Q copies D at the *rising edge* of the clock pulse. Data in the form of logic 0 or 1 can be set up at D, but will only be transferred to Q in step with the rising edge of the clock.

Output Q' always produces the opposite logic level of Q. We will now see how one simple link from Q' to D can produce another vital circuit in electronic processing.

T-type bistable

The T (toggle) bistable is the electronic equivalent of the type of push-button switch used on TVs, etc. Press it once to turn the TV on; press it again and the TV turns off.

Figure 11.5 shows how the output from Q' is used as the data input. The logic level at Q' is copied to Q whenever a clock pulse rising edge is received. In this case, a rising edge is produced by closing switch C.

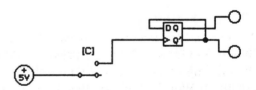

Figure 11.5

Notice that the frequency of output Q is half that of the clock pulse input. This module is sometimes referred to as a frequency divider. Put enough modules in series and you can turn the crystal frequency of 32 768 Hz in an electronic watch into 1 Hz. Figure 11.6 shows how four modules can be cascaded to make a 'divide by 16' counter. Probes have been connected to the Q outputs to avoid confusion. Take advantage of the copy and paste feature of Workbench when drawing this circuit!

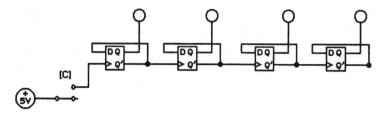

Figure 11.6

Pressing the C switch repeatedly can be tedious, so Figure 11.7 shows how a machine called a 'word generator' can be employed, although here only as a clock generator. Set its output frequency to 1 Hz. In real life, a function generator set to square wave output makes an excellent clock generator.

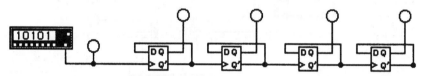

Figure 11.7

Watching the lights on the generator reveals again that each stage of the circuit halves the frequency of the previous stage. It is often helpful to draw a timing diagram to show the relationship between the stages in a circuit. Workbench provides a device called a logic analyzer, which draws graphs automatically. Add the logic analyzer as shown in Figure 11.8. A switch in series with the word generator

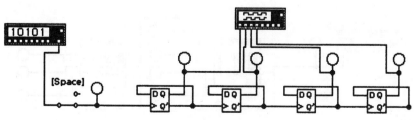

Figure 11.8

controls the point at which the graph starts. Double-click the logic analyzer and set its timebase to 1.00 s/div. Open the switch, click CLEAR on the logic analyzer and close the switch: a graph should appear. The top trace shows output A, the next, output B, etc.

Binary counter

D-type bistables configured as T-type bistables form the heart of binary counters, dividers, shift registers, etc. Not surprisingly it is possible to buy a ready-made IC with T-bistables built in. Workbench offers a binary counter (BIN CRT) IC and this is shown in Figure 11.9. Notice that it has two clock inputs: A and B. Clock A drives only output A; clock B drives outputs B, C, D. To make a

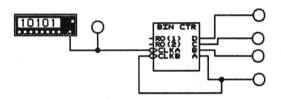

Figure 11.9

four-bit binary counter it is necessary to connect output A to clock B as shown. Note also that the clock inputs have circles indicating negative logic – the clocks respond to a *falling edge*, rather than a rising edge. This is helpful when cascading counters as in Figure 11.10. Try this circuit – using the copy and paste function, noting how the clock input of the second counter is driven from output D of the first. This means that a falling edge is presented to the second counter whenever output D from the first counter, switches from

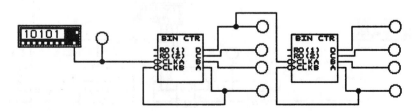

Figure 11.10

logic 1 to logic 0. The whole system could be called a 'divide by 256' counter, as 256 clock pulses cause the final output to change from logic 0 to 1 and back to 0.

Practical note

The counter offered in Workbench is based on the 7493 IC. This may be available as 74LS93, but there appears to be no HC version. There are many alternative counting ICs, such as 74HC393, which is a dual binary counter, i.e. two four-bit binary counters in one IC. Another useful counter from the CMOS 4000 series is type 4029B. This will count up or down in binary or binary coded decimal (explained later). It can also be preset to a particular number if required, which is useful in timers if you wish to count down from 59 seconds, instead of 99 seconds.

Driving seven-segment displays

So far we have used the binary counter to display our count – where 'light on' represents a 1 and 'light off' represents a 0. It is more convenient to display a number in decimal and Workbench includes a display that outputs a decimal number from a binary input. There is no direct equivalent in real life.

Begin by selecting the seven-segment display with *four* inputs. Connect it as shown in Figure 11.11. The display should count through 0, 1, 2, etc. If numbers appear at random, check that the

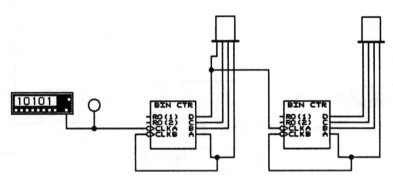

Figure 11.11

connections are in the proper order. If the computer runs rather slowly, try selecting ANALYSIS OPTIONS and change the TOLERANCE to 10% and the POINTS PER CYCLE to 50.

Assuming that a second seven-segment display is also used, note that the units are on the left display, and the tens on the right – this is the opposite of normal counting. Try dragging the displays the other way round – there will be a rat's nest of wires, but Workbench won't mind if you don't!

Try counting beyond nine. Notice that the display does not revert to 0, but displays letters a, b, c, d, e, f. This is because a four-bit binary counter can count to 15 before resetting to 0. The letter 'a' represents 10, b = 11, c = 12, d = 13, e = 14 and f = 15. This type of counting is known as hexadecimal (hex) and it is commonly used when computers are programmed as it is very easy to convert between binary and hex and vice versa.

Binary coded decimal (BCD)

Most people – who do not spend many hours programming computers – prefer to work in decimal. We therefore need to make our binary counter reset after a count of nine, rather than 15. Figure 11.12 shows how this is done.

The binary counter IC can be reset by making both RO1 and RO2 inputs logic 1 at the same time. In other words, the inputs work like the two inputs of an AND gate. Both must be logic 1 for reset to occur. This is useful, as we require a reset only when output D and

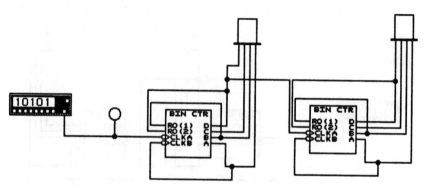

Figure 11.12

B switch to logic 1 at the same time. Remember, output D is decimal eight, and output B is decimal two, so when both switch to logic 1 (decimal 10) we need a reset. If the binary counter did not have this AND input arrangement we could employ an external AND gate to achieve the same result. Try the circuit in Figure 11.12, noting that although the system is working in binary, we have a normal decimal count. This is binary coded decimal.

A real circuit

In practice, a seven-segment display is simply seven LEDs arranged in a figure of eight formation. All the numbers from zero to nine can be displayed by lighting the appropriate LEDs. Quite a complex circuit is required to translate a four-bit binary number into the appropriate code to cause the correct segments to light. However, as always, inexpensive ICs are available that perform the task. The IC offered by Workbench is based on the 7446, 7447, 7448, 7449 decoders (see the note below). It has a four-bit binary input (labelled A, B, C, D), and seven outputs designed to drive the seven segments of the display. The display is a 'common cathode' type, i.e. all the cathodes of the LEDs are joined to one common pin. In real life, this must be connected to 0 V; in Workbench the connection is assumed without being shown.

All decoding ICs have their own peculiarities and a data book (or very good catalogue) should be consulted. The Workbench decoder includes inputs labelled BI (blank input), LT (lamp test) and RBI (ripple blank input). These should all be connected to positive in order to allow the display to work normally. When at 0 V, the BI pin causes the display to blank out, the LT pin causes all segments to light for testing and the RBI pin is used to suppress leading zeros, so that in a four-digit system, the number 0023 would be displayed as 23.

Figure 11.13 shows all the connections; our counter now looks rather complicated and it can only manage up to 99. If six or eight digits are required, the wiring becomes too complicated. A system called 'multiplexing' is employed to allow a whole set of digits to be driven from a single set of outputs. Try viewing an illuminated display, such as the type used on video recorders, then place a revolving fan in your line of sight. The display is not constant as it first appeared; in fact the numbers and letters are 'strobing', i.e. lighting one after another.

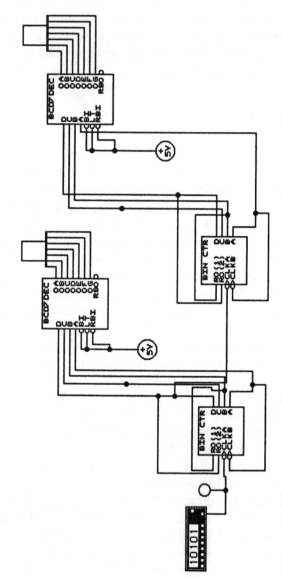

Figure 11.13

Practical note

If building this circuit in real life, a useful seven-segment decoder IC is the 74HC4511 or CMOS 4511B. Be sure to buy a common cathode display and, as always, do not forget to connect up the power supply pins of the IC and one of the common cathode pins on the display to 0 V. The LED segments should be treated like normal LEDs and each requires a series resistor of value, say, 330 Ω.

J-K flip-flop

The J-K bistable or flip-flop can be used in much the same way as the D-type. Figure 11.14 shows a test circuit to help explain the action of this flip-flop. We will refer to the output labelled Q; the output labelled Q′ always does the opposite. Note that, in our test circuit, all the inputs are assumed to be at logic 0 if unconnected. In real life this is not the case and the designer must ensure that the correct logic level is maintained. The set and reset inputs force the output to its set state (Q=logic 1) or reset state regardless of the other inputs or clock pulse.

The J-K inputs control the outputs, but the outputs only change state when a clock pulse is received. The small circle by the clock

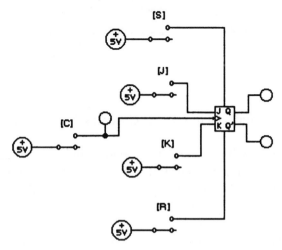

Figure 11.14

pulse input means that the IC responds to a falling edge, so the outputs change state when the switch labelled C is *opened*.

The J-K inputs operate in the following way:

J	K	Q (at falling edge of clock pulse)
0	0	no change
1	0	1
0	1	0
1	1	toggles at each clock pulse

Try closing switches J and K to make J and K logic 1. Notice that the outputs toggle whenever the switch labelled C is opened. The system is behaving like a T-type bistable.

Shift registers

Shift registers are comprised of a group of D-type flip-flops that can store and manipulate data.

1 Try the circuit shown in Figure 11.15. Begin by closing switch D.

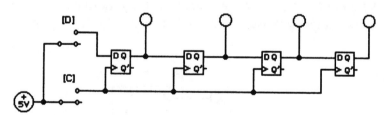

Figure 11.15

2 Then close and open switch C (clock). The left-hand probe should light.
3 Now open switch D. Close then open switch C several times: the next probe to the right will light. If switch D is closed at any time another 'bit' will enter the system and be shifted to the right as switch C is toggled.

A shift register IC is provided by Workbench. Try the test circuit shown in Figure 11.16. Data is set up by means of switches A, B,

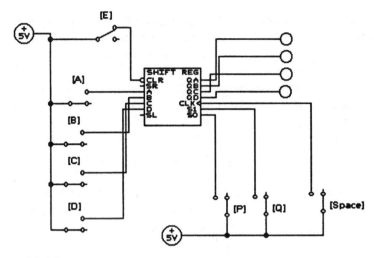

Figure 11.16

C, D. Switch E clears all data when opened. The switches labelled P and Q control inputs S0 and S1 and the switch labelled 'Space' provides a clock pulse when toggled. The outputs (labelled QA, QB, QC, QD) are connected to logic probes. Two further inputs (SL and SR) are unused in this circuit – they will be explained later.

Operation of the circuit is controlled by switches P and Q according to the following table:

P	Q	Operation
0	0	Hold
0	1	Shift left
1	0	Shift right
1	1	Parallel load

Begin by ensuring that switch E is closed. Close switches P and Q to achieve a parallel load. We will load the binary number 0010 by closing switch B. (Remember that binary numbers are written DCBA, so that A is the least significant 'bit'). Press the space bar twice to 'fix' the number into the register. Now open switch Q. This will set the register to 'shift right'. The term 'right' can be confusing, but refers to a shift from A to B to C etc. In Figure 11.16 the lit probe will shift from B to C when the space bar is pressed twice. At the next clock pulse the lit probe will shift down another step.

If switch P is opened and Q is closed, the lit probe will shift up each time a clock pulse is received. Any binary 'word' can be set up on switches ABCD and the whole word can be shifted up or down according to the settings of P and Q. The switches ABCD load the binary word *in parallel*. In other words, all four bits of data are loaded during a single clock pulse. When the data is shifted right, the binary word is shifted bit by bit from output QD. Hence, QD produces a *serial* output. Our circuit is a primitive form of 'parallel to serial converter'.

We can feed serial data in as shown in Figure 11.17. Here, the word generator is employed to feed bits into the serial input (SR).

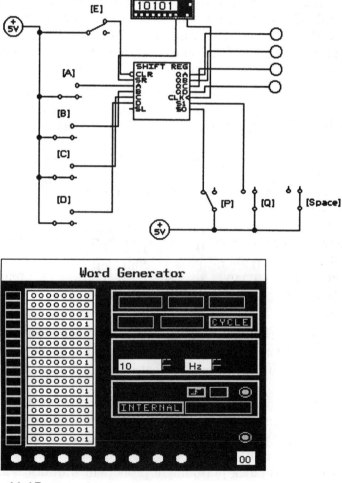

Figure 11.17

We can watch the progress of the data through the probes as the clock is pulsed. Some careful setting up is required:

1 Begin by opening up the word generator and setting vertical pairs of bits to 1 as shown in Figure 11.17. This creates an easy pattern to watch. We will now feed these pairs of bits into the SR input by connecting the right-hand output of the word generator to input SR. A synchronized clock pulse is also required and the word generator provides this.
2 Open switches ABCD (or remove them), but ensure switch E is closed. Switch P must be closed and Q opened to allow the flow of data from A to B to C, etc.
3 Press STEP on the word generator several times to watch the data being fed into the shift register and moving through the outputs QA, QB, QC, QD. Note that a sequence of four clock pulses fully loads the shift register with data. The data could then be extracted from the four outputs if required. Our circuit is a primitive form of 'serial to parallel converter'.

Transmitting data

When two items of equipment are connected – a computer and a printer for example – it is possible to transmit all four bits of data at one time, i.e. in parallel. A four-bit system would require four wires plus a ground wire plus wires carrying control signals. Sending four bits (known as a nibble) in parallel can therefore be expensive, especially over long distances. Sending 8 bits (a byte) in parallel compounds the problem further.

When sending data over long distances, less wires are required if the data is sent bit by bit as in a 'serial link'. Of course it takes longer to send the data – at least four times longer for a nibble; over eight times longer for a byte. Extra circuitry is required: first to convert the parallel data into serial at the transmitter; then back to parallel at the receiver.

Figure 11.18 shows an example circuit that transmits the data fed in by the switches ABCD. It is not a real life circuit, but demonstrates the principle of serial transmission.

1 First clear the system by opening then closing switches E.
2 Close switches P and Q to set the system for parallel data.
3 Set up a binary word at the transmitter by closing one or more of the input switches ABCD.

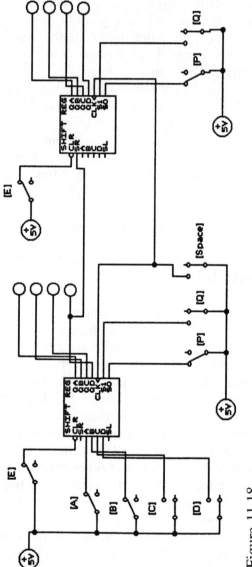

Figure 11.18

4 Clock-in the word by closing, then opening, the clocking switch labelled 'space'.

5 Now open switches Q. This sets the system to serial.

6 Close and open the clocking switch four times. The data shown by the probes should move down the left-hand row and appear on the right-hand row. After four clock pulses the pattern of lights on the right should be identical to the original word.

Notice that all the data passed along a single wire from output QD on the left to input SR on the right. We have cheated a little, first by linking the clock pulse to keep the two circuits exactly in step and, second, by using switches P and Q twice. But the principle should be clear, nevertheless.

Questions

Full written answers, complete with Workbench circuits, are available on the accompanying disk. See p. 224 for details.

1 Draw a D-type bistable, and:

   (a) explain what happens to its output Q, if its set or reset pins are triggered

   (b) explain how the clock input and data input pins work together to affect the state of the output Q

   (c) explain the relationship between output Q and output Q'.

2 You have two clock pulse generators with outputs labelled A and B. These outputs switch rapidly and independently between logic 0 and logic 1. You wish to connect a push switch that 'freezes' the outputs whenever it is pressed, to make a game of chance. Show how this can be done as follows:

   (a) select two clock generators (from the passive parts bin) and set one to 1000 Hz and the other to 500 Hz

   (b) connect the lower side of each clock generator to ground

   (c) connect the top of each clock to dots labelled A and B

   (d) connect a 5 V source to a switch.

   (e) select two D-type bistables: connect a probe labelled A to the Q output of one bistable and a probe labelled B to the Q output of the other

   (f) complete the circuit to achieve the purpose outlined earlier.

3 (a) Show how a D-type bistable can be connected to make a T-type bistable.
  (b) Explain what happens to output Q when a series of pulses are fed to the clock input.
  (c) What is the frequency of the output compared with the input?

4 Design an up-counter circuit based on three D-type bistables (made into T-types) with set and reset inputs. Connect probes to the Q outputs of the bistables. Connect the clock input of the first bistable to a switch operated with the space bar. Connect all the set inputs via a single NOT gate to a switch operated by letter S. Connect all the reset inputs via a single NOT gate to a switch labelled R. Connect all three switches to a 5 V source.

  (a) Reset the system by closing then opening switch R.
  (b) How many times must the space bar switch be closed to make the last probe light?
  (c) How many times must the space bar switch be closed to make the last probe switch on then off (assuming the circuit is first reset).
      Retain your circuit for the next question.

5 Modify your circuit in Q. 4 so that the system counts down rather than up. Retain your circuit for the next question.

6 (a) Add a seven-segment display (the four-binary input version) to your circuit in Q. 5, so that it correctly registers the binary number represented by the probes.
  (b) Convert the following numbers into hexadecimal:
      (i)   5
      (ii) 10
      (iii) 14
  (c) Explain the term 'binary coded decimal'

7 Design a circuit based on J-K flip-flops, which divides by 32.

8 (a) Show how four bistables can be wired to make a shift register. Connect all the clocks to 5 V via a switch C. Connect the left-hand data input to 5 V via a switch S (for serial). Check that the system works.
  (b) Now add four switches 1, 2, 3, 4 so that parallel data can also be inputted to the system. You will also need three diodes to make the circuit work successfully.

Chapter 12

Adders, comparators and multiplexers

When designing a circuit that will add binary numbers together, it is helpful to analyse exactly what is required. Our system must produce the following results:

$$0 + 0 = 0$$
$$0 + 1 = 1$$
$$1 + 0 = 1$$
$$1 + 1 = 0 \text{ and carry } 1$$

This can be written in the form of a truth table with inputs A and B:

A	B	SUM	CARRY
0	0	0	0
0	1	1	0
1	0	1	0
1	1	0	1

A glance at the truth tables discussed in Chapter 7 shows that the sum results are obtained with an exclusive OR gate, i.e. the outputs are at logic 1 when A OR B *but not both* are at logic 1. The carry output is simply the result of an AND gate, i.e. A AND B must be at logic 1 for the output to be at logic 1.

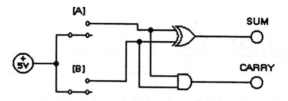

Figure 12.1

Half adder

The circuit that achieves this is shown in Figure 12.1. Try the circuit, noting the result shown by the probes. The system is known as a 'half adder'. It can only provide the result of two binary inputs. Workbench provides a ready-made half adder, and this is shown in Figure 12.2. Try the circuit and verify that it provides the same result. Note that the outputs are labelled Σ (meaning sum) and Co (meaning carry out).

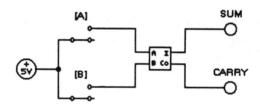

Figure 12.2

The term 'half adder' implies that this is not the whole story. Try manually adding two four-bit binary words together,

$$\text{e.g.} \quad \begin{array}{r} 1111 \\ + 0110 \\ \hline 10101 \end{array}$$

Starting at the right-hand side (i.e. the least significant bit), 1 + 0 =1. In the next column, we have 1 + 1 = 0 and carry 1. So far we have only required half adders, as only two digits have been added. However, in the next column we have 1 + 1 plus the 1 carried from the previous column, i.e. 1 + 1 + 1 = 11. Our half adder cannot add three bits, as there are only two inputs: A and B.

It follows that, when two binary words are added, any of the columns could include three bits except the least significant, i.e. the

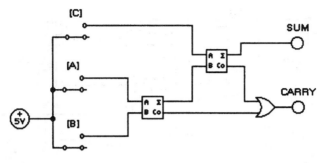

Figure 12.3

column on the right, as any column could produce a carry bit. So our half adder is only adequate for the right-hand column.

Full adder

A full adder must be able to add three bits and a suitable circuit is shown in Figure 12.3. Here switch C simulates 'carry in'. Try the circuit – it correctly adds the result of the settings on switches A, B, C. A full adder is provided by Workbench and Figure 12.4 shows how it is connected.

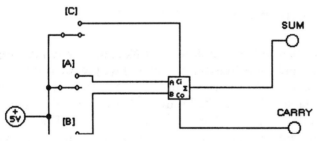

Figure 12.4

We are now in a position to create a circuit capable of adding our two binary words shown earlier, i.e. 1111 + 0110. Figure 12.5 shows how it is done. The switches A, B, C, D are used to set the first binary word, with switch A being the least significant, and switches E, F, G, H set the second word. The probes show the result – notice that five

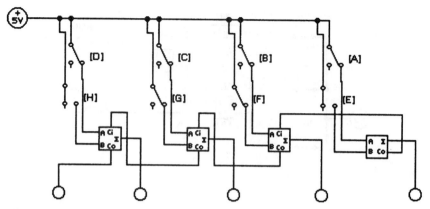

Figure 12.5

are required, as two 4-bit numbers can result in a 5-bit number when added. The result of $1111 + 0110$ should be 10101.

The system shown in Figure 12.5 is a parallel adder, i.e. it processes the two binary words in a single step. It is possible to add two binary words using a serial adder; the saving is in the amount of circuitry required, but it takes longer to process the data.

In practice, all the circuit modules described, including complete binary adders, can be purchased in the form of ICs.

Binary comparator

It is often useful to be able to compare the values of two binary words and indicate if they are equal. The circuit shown in Figure 12.6 compares the nibbles set up on two groups of switches. If the switches labelled A, B, C, D are set in exactly the same way as those labelled 1, 2, 3, 4, then the probe will light. The exclusive OR gates compare the settings of each pair of switches (A with 1, B with 2, etc.) and output a logic 0 if they are the same. If all four exclusive OR gates output a logic 0, then the 4-input NOR gate outputs a logic 1 indicating a match.

Figure 12.7 shows how the circuit may be extended to make a simple combination entry system. The four switches labelled 1, 2, 3, 4 would be hidden toggle switches used by the operator to set up the correct entry code. The five switches labelled A, B, C, D and E would be momentary types. The user would press the correct combination, then press E for entry.

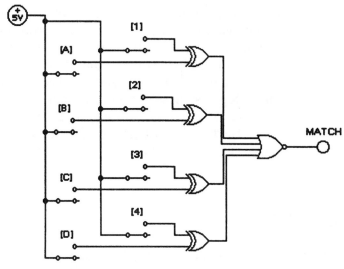

Figure 12.6

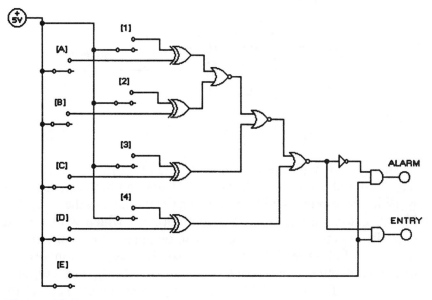

Figure 12.7

If the correct combination is chosen, the *entry* probe lights when E is pressed. If the combination is wrong then the *alarm* probe lights. In practice, there may have to be a latching circuit to keep the alarm sounding if switch E is released, or a timer (monostable) circuit from Chapter 9. Notice how the 4-input NOR gate has been constructed from three 2-input NOR gates; these perform an identical task and, as four are available in an IC such as 74HC02 or CMOS 4001 the cost of the circuit is reduced. The fourth NOR gate could be used to make the NOT gate by joining its inputs together.

Multiplexers

Multiplexing is an essential part of electronics and the circuits described should help clarify the principles involved. Figure 12.8 is a test circuit based around the '1 of 8 multiplexer' (1–8 MUX).

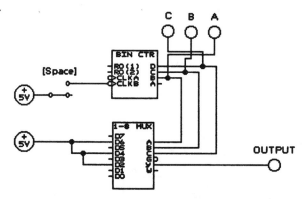

Figure 12.8

This type of multiplexer can be likened to a rotary switch with one pole and eight 'ways'. This particular multiplexer can only work in one direction with the pole as the output and the 'ways' as inputs. Later, we will look at a demultiplexer which works in the opposite direction. It is also possible to obtain ICs that work in both directions. In Figure 12.8, the multiplexer has eight data inputs (D0 to D7) and three binary control inputs (A, B, C). The output is labelled 'y'. The job of the IC is to connect the output to one of the data inputs; it selects which input according to the binary code at A,

B, C. For example, the binary code 010 is equal to decimal two. So, if A and C are held at logic 0 and B at logic 1, the output of the IC will be connected to data input D2. Hence, the logic level presented to D2 will be copied to the output.

The binary inputs A, B, C can be set from 000 (decimal 0) to 111 (decimal 7). If we employ a binary counter, as shown in Figure 12.8, we can 'clock' the counter and cause the multiplexer to scan through all its data inputs. In the circuit we have made the even data inputs D2, D4, D6 equal to logic 1. When the space bar is toggled, the probes labelled A, B, C will show the binary number being fed to the multiplexer. When binary 010 (2) is reached the output probe should light; likewise with binary 100 (4) and binary 110 (6). The output labelled 'w' produces the opposite logic level to output y. Input G causes output y to always be at logic 0 if G is at logic 1.

Demultiplexer

Figure 12.9 shows how the data encoded by the first IC may be decoded. It is necessary to connect the inputs A, B, C to the counter so that the demultiplexer remains in step with the multiplexer. The demultiplexer (based on IC type 74155) has active low inputs; hence the NOT gate is required to invert the logic level.

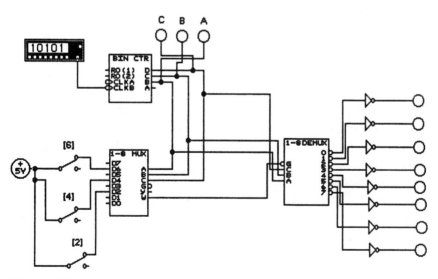

Figure 12.9

The word generator should be set to cycle. When the counter clocks, the probes connected to the outputs 0 to 7 will light one by one according to the data on the input side. As we have made inputs D2, D4 and D6 logic 1, probes 2, 4 and 6 are the only ones that will light. Hence, any data set on the input side will be copied to the right-hand side. Eight bits of data are transferred, yet only four wires (plus a ground, which is not shown) are needed. However, each probe on the right-hand side has a 1 in 8 'look' at the data on the left.

The usefulness of the system depends upon the speed of the binary counter. Set the word generator in Figure 12.9 to cycle and its running speed to 10 Hz or more. The data inputs D2, D4, D6 have been connected via switches. Ideally, *all* the data inputs would be connected to 5 V via switches; this may be possible to demonstrate with a large-screen computer. When the system is running, it should be possible to copy any switch settings to the probes on the right-hand side.

As no two probes can be lit at once, the system will have to be run at a sufficiently high speed for the result will appear constant. You will need a super-fast computer, or a little imagination, to appreciate this. Virtually all electronic displays are strobed in a similar way, but their operating speeds are so high that we are not normally aware of this.

Chaser

The circuit in Figure 12.10 shows how a demultiplexer can be made into a simple light chaser. The chasing speed is controlled by the clock speed setting on the word generator. In practice, one of the astable modules provided in Chapters 9 or 10 could be employed. Many binary counters, such as the CMOS 4029B, are able to count up or down, and so the chasing direction can be reversed.

It is possible to obtain demultiplexers that do not require the NOT gates shown in the diagram. Alternatively the outputs could be employed as current sinks, for example, turning on a pnp transistor or pnp darlington. The pin labelled 'G' (which can be likened to the pole of a rotary switch) is rather misleading; remember that Workbench assumes that unconnected inputs are at logic 0 (i.e. connected to 0 V) and, as the input is inverting, no connection is required. A 'normal' input would have to be connected to positive.

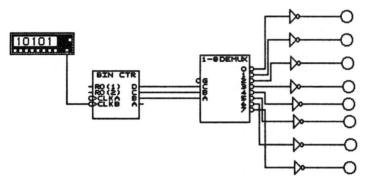

Figure 12.10

Questions

Full written answers, complete with Workbench circuits, are available on disk – see back of book for details

1 (a) Show how an exclusive OR gate and an AND gate may be used to add two inputs A and B and provide a sum and carry output. Employ two switches, so that your inputs can be made logic 0 or logic 1, and two probes labelled 'sum' and 'carry'.
  (b) Why is this circuit called a 'half adder'?
  Retain your circuit for the next question.

2 (a) Show how a full adder is achieved using only logic gates (i.e. not the half adder or full adder provided by Workbench). Use three switches A, B, C, where switches A and B represent the binary bits that must be added together, and C represents the carry input. Use two probes labelled 'sum' and 'carry'.
  (b) What happens at the sum and carry outputs when all three switches are closed?

3 Two 2-bit binary numbers are to be compared. If they are the same, a buzzer must sound. Design the circuit and set the buzzer to 4 V, 0.05 A. You should power the circuit with a 5 V battery, not forgetting a ground symbol. The binary numbers should be set with switches A and B, and J and K, so that, if the setting of A = J and B = K, the buzzer sounds.

4 Show how a demultiplexer can be used to make a light chaser system that resets automatically after the third probe has lit. Base your design on the circuit shown in Figure 12.10 and remember that the binary counter resets when both inputs RO(1) and RO(2) are made logic 1 together.

Chapter 13

Field effect transistors

The field effect transistor (FET) has some very useful properties, but is more difficult to understand than the ordinary bipolar transistor. There are several varieties of field effect transistor, but the common ones are shown in Figure 13.1. Each is a 'junction gate FET' (JFET or JUGFET). We will start with the n-channel type. Note that, like an ordinary bipolar transistor, there are three terminals, but here they are called gate (G), drain (D) and source (S).

n-channel JFET   p-channel JFET

Figure 13.1

The test circuit shown in Figure 13.2 allows us to examine some of the properties of the JFET. Try the circuit and double-click the potentiometer so that its increment can be set to 1 % – this allows fine control over the voltage applied to the gate. The arrangement of two batteries allows us to vary the gate over a wide range of voltages, positive and negative.

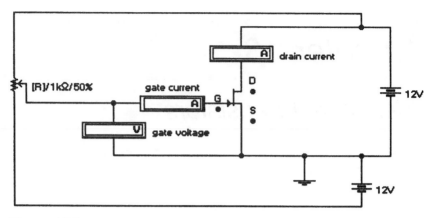

Figure 13.2

We will begin by destroying the JFET – at least, a real one would be destroyed! Try reducing the potentiometer setting to less than 50 % so that the gate voltage rises above zero. *This should never be done in real life.* The drain current rises, but then stops rising at a safe 22 mA, or so. However, the gate current rises to an alarming extent, peaking at over 1000 A as the pot is reduced to 0 %.

The gate/source junction behaves like a diode, rather like the base/emitter junction of a bipolar transistor. But, unlike the bipolar transistor, we must never make the gate more positive than the source, i.e. *current must never flow into the gate.* The JFET is controlled by making the voltage at the gate *lower* than the voltage at the source. Hence, the diode inside is reverse biased. Current will not flow out of the gate, but a field is created that affects the behaviour of the JFET.

Set the potentiometer to 50 %. The gate voltage and gate current should both be zero and the drain current should be about 12 mA. Now increase the setting above 50 % in steps of 1 %, so making the gate voltage negative. The drain current will decrease, and, when the gate voltage crosses the –2 V point, the drain current becomes zero. Note that the gate current remains at zero – remember that the diode inside the JFET is reverse biased, so no current will flow. The figures of 12 mA and –2 V apply to the standard JFET in Workbench. In real life, these figures are likely to be different – they even vary from one JFET to another of the same type.

The fact that there is no gate current is one reason why JFETs are so useful. With a gate current of zero, the JFET is a 'voltage controlled device'. Ordinary bipolar transistors require current into the base. The JFET can be controlled without using current from the previous stage of the circuit; we say that it has a very high input impedance. We have already seen how CMOS logic gates make circuit design easy due to the very high impedance of their inputs. In fact, CMOS gates are based on another type of FET.

The FET as a variable resistor

The last circuit demonstrated how the JFET can be used to control the flow of current into the drain terminal. The JFET is effectively a variable resistor and we can calculate its resistance when the gate voltage is zero (i.e. R set to 50 %). As the JFET is the only device in series with the 12 V battery (apart from the ammeter, which we assume has zero resistance), we can use Ohm's Law to calculate the resistance of the JFET. The drain current is 12 mA (0.012 A), so:

$$R = 12/0.012 = 1000\,\Omega$$

If we reduce the gate voltage still further (i.e. make it more negative) then the current flowing from gate to source falls to zero. So our JFET can be set to any resistance between 1 kΩ and infinity.

The FET as an analogue switch

The variable resistor example shows that the JFET can be made to have infinite resistance, i.e. an open circuit. The circuit shown in Figure 13.3 shows how the JFET can be employed to control an analogue signal from the function generator. Begin by setting the function generator to sine wave, with an amplitude of 2 V. The oscilloscope will show the action of the system. The purpose is to allow a logic level to switch on or switch off the signal through the JFET. In this case, the logic level is set by means of a toggle switch. Try the circuit: the wave trace should appear on the oscilloscope screen. Now close the switch so that *negative* 5 V is applied to the diode. The wave trace should now be a straight line, showing that the signal cannot pass through the JFET.

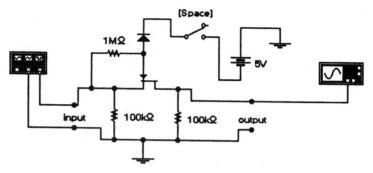

Figure 13.3

Although the labels 'input' and 'output' are shown in Figure 13.3, in fact either side can be an input or an output. Try interchanging the connections from the signal generator and oscilloscope: the system should work as before. We have created a bidirectional switch – a switch that can be controlled by a logic level and which will allow an analogue signal to pass in either direction.

There are many applications for this type of system; perhaps the most obvious is an amplifier that can be controlled via a remote handset. When the amplifier is changed from, say, 'CD' to 'tape' by means of the handset, the signals are often routed via analogue switches made from FETs. In practice, a more advanced type of FET is employed and ICs are available with all the required circuitry. The most obvious drawbacks of the circuit in Figure 13.3 are that the analogue signal cannot have an amplitude greater than 2 V and a –5 V supply is required. The latter problem is easily cured using a p-type JFET, rather than the n-type we have used so far. The

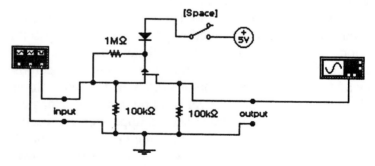

Figure 13.4

p-type works in a similar way to the n-type, except that all the polarities are reversed. Figure 13.4 shows how the p-type can be employed to make an analogue switch. This time, we simply require a +5 V supply to control the switch. Try the circuit, noting that, when the toggle switch is open, the signal can pass through the JFET.

Constant current sink

The circuit shown in Figure 13.5 illustrates the way in which the JFET can be employed as a constant current sink. The ammeter in the circuit represents the load. Try the circuit and note the reading on the ammeter.

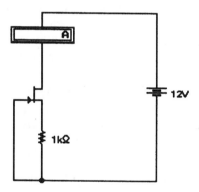

Figure 13.5

Now change the voltage of the battery. The reading on the ammeter will not change, in spite of large variations in battery voltage. Even if the resistance of the load is changed the current will still be unaffected. As an example, try a resistor of 1 kΩ in series with the ammeter. Of course, if the value of the resistor is raised so high that the battery voltage is insufficient to maintain the present conditions, then the reading on the ammeter will fall.

The only problem with this circuit is that different JFETs, even of the same type, will have different parameters (i.e. factors which affect their performance). In real life it is not possible to accurately

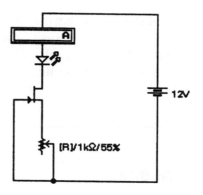

Figure 13.6

predict the sink current. The solution is to use a variable resistor (such as a preset) and Figure 13.6 shows an example where a low current LED (e.g. an LED that requires about 2 mA) is being driven via the JFET.

This circuit is useful if the battery voltage is unpredictable, e.g. you may wish to use a control that varies the speed of a motor by changing the voltage across it. If you also required an LED wired across the motor connections, it would be difficult to calculate the value of the series resistor required, as the supply voltage is variable. A JFET solves the problem by providing a fixed current regardless of the supply voltage (providing it does not fall below 3 V).

Try the circuit in Figure 13.6 and verify that, no matter how much the battery voltage is increased, the current through the LED does not rise above 2.1 mA when the variable resistor is set to 55 %.

AC amplifier

The circuit in Figure 13.7 shows how a simple JFET amplifier may be constructed. Try the circuit and set the frequency of the function generator to 1 kHz and the amplitude to 50 mV. This signal level resembles the signal produced by a typical microphone. Set the oscilloscope timebase to 0.20 ms/div and both input channels to 200 mV/div.

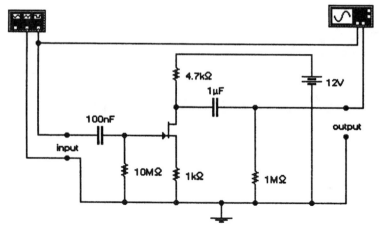

Figure 13.7

Questions

Full written answers, complete with Workbench circuits, are available on disk – see back of book for details

1 (a) Show how a JFET can be connected to control the brightness of a bulb that is set to 12 V, 0.01 W. Use two 12 V batteries and a 10 kΩ potentiometer. Place an ammeter in series with the bulb, so that the current can be monitored. Add a resistor, so that the gate voltage cannot rise above 0 V regardless of the setting of the potentiometer.
   (b) Why is it important that the voltage at the gate of the JFET is not allowed to rise above 0 V?
   (c) Why is the JFET referred to as a voltage controlled device?

2 Draw a JFET circuit that can switch an AC voltage source on and off. Use a bulb connected to the output. Set the AC source to 1 V, and set the bulb to 1.5 V, 0.001 W.

3 Show how a JFET can be used to supply a constant current of about 6 mA through an LED regardless of changes in voltage from the power supply. Test your circuit with a single supply of 12 V to 50 V.

Chapter 14

Thyristors, triacs and diacs

The silicon controlled rectifier (SCR), or thyristor (its common name), is used mainly with AC circuits and performs a similar task to a transistor, except that it generally works at high (mains) voltages. It is for this reason that experimental work is best avoided, but we will look at a possible low voltage application.

Latching circuit

The circuit in Figure 14.1 illustrates the function of a thyristor. The three terminals are known as anode, cathode and gate. Like a diode, current will not flow from cathode to anode, but current will flow from anode to cathode if a small voltage is applied to the gate. Once the current is flowing, it will remain until interrupted in some other way.

1 Try the circuit. Begin by closing switch B.
2 Now close switch A – the bulb should light.
3 Open switch A – the bulb should remain on, showing that the thyristor is now latched on.
4 Open, then close, switch B. The interruption to the supply should have made the thyristor unlatch.

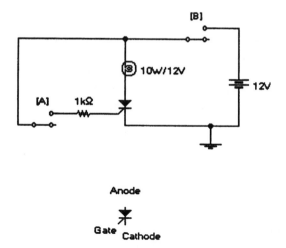

Figure 14.1

Simple alarm circuit

A simple latching alarm circuit can be made as shown in Figure 14.2. Care must be taken to prevent the thyristor being triggered accidentally, and so the 2.2 kΩ resistor is included to ensure that the gate voltage is zero unless the alarm trigger switch (A) is closed. The capacitor removes any voltage spikes that may find their way into the input, especially if switch A is connected via long wires. Switch A could be an under-carpet pressure switch and switch B (master on/off) could be a key-operated switch.

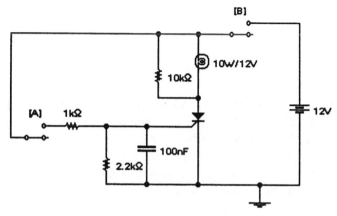

Figure 14.2

Try the circuit, and replace the bulb with a buzzer to represent the siren. A 10 kΩ resistor is shown in parallel with the bulb in Figure 14.2. In practice, this is unnecessary if a bulb or relay is used; however, some buzzers and sirens cause interruptions to the flow of current and these can cause the thyristor to unlatch accidentally. The 10 kΩ provides a path, so that there is always a flow of current.

AC control

The thyristor can be used for AC control just as the transistor is used for DC control. The thyristor will not latch on, as the current flowing from anode to cathode stops each time that the AC supply reverses. Figure 14.3 shows an example circuit. Workbench cannot simulate quickly enough to show the true effect of the circuit and some interpretation is needed.

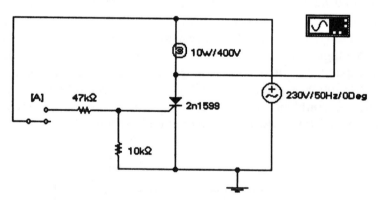

Figure 14.3

Note that the default thyristor cannot cope with the UK mains supply of 230 V RMS (325 V peak). Hence, thyristor type 2N1599 has been selected from the list provided, to handle a peak of 400 V. A popular and inexpensive thyristor available in the UK is type C106D – this can handle a peak of 400 V and has a maximum current rating of 4 A. The bulb voltage has been raised to 400 V; in real life the rating of a bulb is taken as the RMS figure.

Set the oscilloscope timebase to 5.00 ms/div and the input to 100 V/div. When switch A is open, the bulb should not light and the waveform will be a normal AC graph. In other words the same voltage will be present on both sides of the bulb throughout each AC mains cycle.

Now close switch A. The bulb should light for each half-cycle when the thyristor turns on. However, if switch A is opened, the bulb should switch off and remain off.

In real life, the bulb would flash so quickly that it would appear to be lit continuously. However, as current is only conducted through the thyristor in one direction, the bulb will appear to be only 'half' bright. We could use two thyristors – one pointing in each direction – to cure this problem, or a device known as a 'triac'.

Note: the circuit shown in Figure 14.3 is not intended for use in real life, and experiments involving direct connection with the mains supply should only be attempted by a suitably qualified person.

Triac control

A triac is essentially a pair of thyristors wired in inverse parallel, with a single controlling gate. The two ends of the triac are called T1 and T2. The circuit shown in Figure 14.4 illustrates the operation of a triac. Ensure that type 2N5568 is selected from the list available. In the UK, type C206D provides a peak of 400 V and a current capacity of 4 A. The power of the bulb should be increased to 100W to provide a reasonable current through the triac.

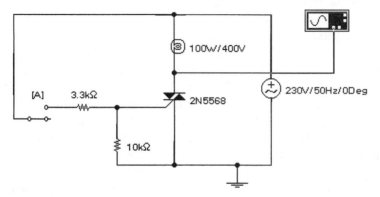

Figure 14.4

Try the circuit using the same oscilloscope settings as before. When the switch is open, the bulb should remain off and the oscilloscope waveform is identical both sides of the bulb. When the switch is closed, the bulb will light whenever the voltage at the gate exceeds a certain minimum in the positive and negative direction. This figure depends upon the type of triac employed.

The wave trace on the oscilloscope may be quite confusing and a more logical result can be obtained if the bulb is placed below the triac. The oscilloscope is then connected at the top side of the bulb. This method employed in the next circuit.

If the system could simulate in real time, the bulb would appear to be fully lit. The switch could be a second triac – perhaps an optically triggered type, where the light from an LED causes the triac to conduct. This type of device (known as an opto-isolator) is useful if you require a sensitive and expensive device, such as a computer, to switch mains lights on and off. The computer merely controls an LED; there is no connection between the computer and the mains supply.

Note: the circuit shown in Figure 14.4 is not intended for use in real life, and experiments involving direct connection with the mains supply should only be attempted by a suitably qualified person.

Lamp dimmer

The circuits shown so far only enable the bulb to be switched on and off. Our final triac circuit allows the bulb to be dimmed. In order to achieve a more understandable wave trace the bulb is placed below the triac in Figure 14.5. A new component, known as a diac, is also employed. A diac conducts in both directions, but only when the voltage across it exceeds its switching voltage. This value depends upon the type of diac employed.

During the positive part of the AC cycle, the capacitor charges and, when the voltage across it exceeds the switching voltage of the diac, a rush of current flows through the diac and into the gate of the triac. The triac is turned on and the bulb lights. If the variable resistor is set to a high resistance (say 55 %), it takes nearly a half-cycle for the capacitor to charge sufficiently, hence the bulb lights for a very short time. If the variable resistor is set to a lower resistance, the triac is turned on earlier in the AC

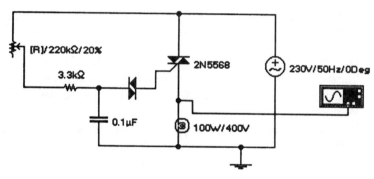

Figure 14.5

cycle, so the bulb is on for longer. Exactly the same happens during the AC negative half-cycle. The net effect is that the lamp appears to be dim or bright, according to the setting of the variable resistor.

Clearly, Workbench cannot run fast enough to enable the bulb to appear to change in brightness and a little imagination is required. However, the graph on the oscilloscope shows exactly what would happen in real life.

This method of brightness control is in common use, from dimmer switches in homes to theatre lighting control systems. Professional circuits will include devices to reduce transients – high voltage spikes produced by the rapid switching of the current.

Note: the circuit shown in Figure 14.5 should not be constructed in real life; there are a number of safety features not included, and no components have been included to reduce transients and radio interference.

Questions

Full written answers, complete with Workbench circuits, are available on the accompanying disk. See p. 224 for details.

1 A typical 'steady hand tester' as used at fêtes, consists of a wire loop that has to be passed along a wire without touching it. If you do touch the wires, then a buzzer sounds, indicating that you have failed. The problem is that a slight touch of the wires will

cause the buzzer to sound for such a short time that it could be difficult for the operator to hear.

(a) Design a circuit based on a thyristor, which will latch the buzzer on when the two wires touch – even if they touch for a very short time. Simulate the steady hand wires by a switch W and include a master on/off switch M.

(b) Include any additional components you think necessary that will prevent the buzzer sounding accidentally, even if electrical interference is picked up by the wires.

2 (a) Design the simplest circuit that shows a thyristor employed to switch a lamp on or off. Use a 12 V AC supply, and a 20 V lamp. Control the thyristor with a switch.

(b) When the switch is closed, why does the bulb appear to flash on and off?

(c) If the switch is opened just after the bulb has lit, there is a delay before the bulb switches off. What causes the delay?

3 (a) Design the simplest circuit that employs a triac to control a lamp. Again, use a 12 V AC voltage source, a 20 V lamp and control the triac with a switch.

(b) What is the main difference between the action of this circuit and that of the previous one?

Chapter 15

Constructing your circuit

A circuit design has been perfected with the help of Workbench. How do you make it into a practical circuit? Look once more at the circuit diagram. Does it do exactly what is required? Are all the components annotated (e.g. R1, R2, R3, etc.)? Are all the components labelled with their values? Have you specified which type of IC and which type of transistor is to be employed? You can save a great deal of time by attending to these points at this stage, rather than when the circuit is partly built.

Decoupling

Decoupling capacitors have been omitted in most of the circuits in this book. This is because they perform no useful function in a Workbench diagram and tend to slow down the simulation. In real life, they should be included, particularly in sensitive circuits. Decoupling capacitors are connected across the supply rails as shown in Figure 15.1. Their purpose is to smooth out power irregularities in the supply rails.

In general, use a large electrolytic capacitor – a value of $1000\,\mu F$ will be sufficient in most cases. Select a larger value if a current-hungry device, such as a solenoid, is required at the output; select a smaller value (say $470\,\mu F$) if the circuit is only required to drive LEDs or small buzzers. If in doubt, a larger value will do no harm.

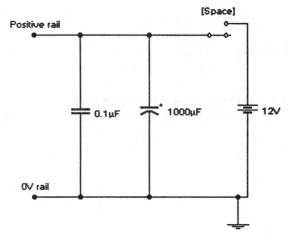

Figure 15.1

Large electrolytic capacitors are inefficient at removing very short voltage spikes, so a ceramic disc capacitor of about 0.1 µF (100 nF) should also be connected across the supply rails. The small capacitor will be most effective if placed physically close to the most sensitive part of the circuit – often the IC. It is wise to include more than one such capacitor in very large designs, so that single ICs or groups of ICs have their own decoupling capacitor.

Add the decoupling capacitors to your circuit design, plus any other safety precautions such as fuses, switches and indicator lights. If LEDs are used as indicators, do not forget that normal LEDs require series resistors (see Chapter 2). Alternatively, use flashing LEDs, or LEDs with built-in resistors, neither of which require a series resistor.

Power supply

You should have chosen your supply voltage by now, but it is surprising how often this is overlooked. The supply voltage is often decided by the type of IC employed, and details can be found in the appropriate chapter. If you have not needed an IC, then decide the supply voltage according to convenience. A small PP3 battery will supply 9 V and will produce enough current to light a small number of LEDs for an appreciable time. A set of AA cells can be used to produce any required voltage; three AA cells in series will produce 4.5 V – ideal for use with CMOS 74HC ICs. Eight AA cells in series will produce 12 V – useful if an output solenoid is required, but any logic ICs in your circuit must now be from the CMOS 4000 family. A 741 op-amp will operate from a single rail supply of between 6 V and 30 V.

Most of the current used by your circuit will be required at the output stage. It is wise to have some idea of the current required as this will affect your choice of battery or power supply. Most catalogues now provide information regarding how much current a particular battery or cell can source. The information is presented in amp hours (AH) or milliamp hours (mA-H). So, a battery rated at 1 AH can provide one amp for one hour, or 2 A for half an hour, or 0.5 A for two hours, etc.

Having chosen your power supply, recheck your circuit to ensure that the supply is adequate. For example, you may have opted for a supply of 4.5 V, but overlooked the fact that this is hardly enough to power two LEDs in series. In this case, it would be a simple matter to operate the LEDs in parallel. The situation would be more serious where your circuit – based on 74HC ICs for example – is required to operate a 12 V solenoid. You might consider a relay with a 5 V coil; alternatively a transistor could be employed to act an interface between the two voltages. Figure 15.2 shows how a transistor could be used. A darlington pair (e.g. TIP121 or TIP122) is suggested, as a large gain and large output current are required. You could employ two power supplies, but Figure 15.2 shows how a regulator IC (type 78L05) is used to provide the 5 V supply for the

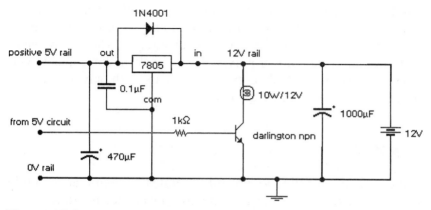

Figure 15.2

logic gates. A primitive version of this IC was created in Chapter 4 and called type 7805. The real life 7805 can handle a current of up to 1 A. However, most CMOS logic circuits will require much less than this, so a type 78L05 is suggested, which can handle a maximum of 100 mA. Note that the input lead (in) and output lead (out) have been interchanged. Note also the extra decoupling capacitors required and the 1N4001 diode across the regulator. The diode may appear to be redundant, but it prevents the voltage at the output ever being greater than the voltage at the input; such an event would destroy the IC.

To summarize, careful thought is needed regarding the choice of voltage. Workbench makes things very easy, as key components, such as relays and bulbs, can be double-clicked to set their voltage and current requirements to any level.

Construction method

The circuit diagram has been checked and double-checked, so we are at last ready to construct the circuit There are a number of construction methods available: two will be mentioned in brief and a third outlined in detail.

Prototype board

Prototype board (often called breadboard) allows a circuit to be built very quickly, as the leads of components (including ICs) are simply pushed into place. It is rather easy to make mistakes, but prototype board is ideal if you wish to modify your circuit, or try it out, then use the parts to make another circuit. When component leads are inserted into prototype board they make contact with metal conductors, which connect the lead with any other lead pushed into the same row of holes.

Stripboard

Stripboard (often called veroboard) operates on a similar principle to prototype board, except that the component leads are soldered into place. Stripboard is much less expensive than prototype board, but it can only be used once. If the leads of two components have to be connected, they are inserted into the same row of holes, which are joined by a copper track. A stripboard circuit was shown in Figure 8.7(b). It is sometimes necessary to break a track to prevent electricity flowing to the wrong place, and a track cutting tool is available. For example in Figure 8.7(b) the stripboard tracks must be broken underneath the IC to prevent the left-hand row of IC pins from connecting with the right-hand row.

Components are usually placed on the non-copper side of the stripboard and their leads may be easily soldered to the strips on the other side. Stripboard provides a fast, convenient method of circuit construction. However, it is very easy to make mistakes and so is not suitable for mass production.

Printed circuit board

Printed circuit board (PCB) is the favoured method of circuit construction. A well-designed PCB is easy to construct and ideal for mass production. A PCB must be designed for a particular circuit – last minute changes are possible, but should be avoided by carefully checking the circuit and layout diagrams first.

Printed circuit boards feature a pattern of copper tracks, which join together the leads of components as required. Single-sided boards (i.e. boards with a single layer of copper) generally have the components on one side, and the copper tracks on the other. The leads of each component are pushed through holes which must be drilled through the board. The leads are soldered to copper pads, and the pads are connected by means of copper tracks.

A PCB is generally produced from a copper-clad board, i.e. a base material, such as fibreglass, coated on one side with copper. The required layout of pads and tracks is produced by etching (dissolving) the copper areas not required. The chemical used to dissolve the copper is known as an etchant and the most common etchant is ferric chloride. Ferric chloride is rather messy, stains badly and must be handled with care. A cleaner alternative is 'clear etchant': this works just like ferric chloride, but does not stain. However, it is also dangerous and must be treated with respect.

When etching a PCB, the process is much more efficient if the etchant is warm and air is pumped through the solution. Although a photographic developing dish could be used, a much safer, if expensive, alternative is the use of a PCB etching tank. This will maintain the liquid at the correct temperature and pump air as required. It is very important to wash the board thoroughly when the etching process is complete.

Designing a PCB layout

Remember that, with single-sided PCBs, the components are placed on the non-copper side, and the leads of components are soldered to pads on the copper side. Begin by sketching your layout using a pen to draw the components and a pencil to show the pads and tracks.

   We will take the example of a simple transistor circuit, which makes a buzzer sound when an LDR is shaded from light. Our circuit will be based on Figure 3.2, except that the bulb will be replaced by a buzzer. Figure 15.3 shows our final circuit diagram with decoupling capacitors added and a 9 V battery. The lower end of the variable resistor (R) has been connected to the wiper. Although this does not make any difference to the operation of the variable resistor, it is good practice and prevents an open circuit occurring if dust causes the wiper to lose contact when it is rotated.

   The BC184 transistor has a gain of at least 200, so the base resistor has been increased in value to 4.7 kΩ. This provides a current of $(9 - 0.7)/4.7\,k = 1.77\,mA$ into the base of the transistor.

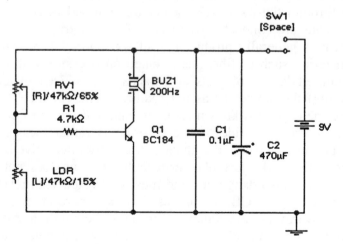

Figure 15.3

Assuming a gain of 200, the collector current could be a maximum of $1.77 \times 200 = 354\,\text{mA}$. This is much larger than required and our choice of resistor will ensure that the transistor is fully saturated. When the transistor is saturated, the *actual* collector current (which should not be allowed to exceed 200 mA) will be determined by the buzzer. Set the buzzer to 6 V, 0.05 A for best results. (When selecting a real life buzzer, choose a 6 V 'solid state' or 'electronic' type.)

When designing a PCB, you will need to be aware of the size of components and the positions of pins or leads on components that must be connected a certain way round, such as on diodes, electrolytic (polarized) capacitors, transistors and ICs. A good catalogue will provide much of this information. Failing that, manufacturers data sheets are available and some suppliers produce their catalogues on CD ROM, in which case data sheets will be included. Note that transistors are almost always drawn with their leads pointing towards you – but when they are fitted into a PCB the leads face away from you.

A PCB layout can look like a circuit diagram and a beginner to PCB design might start with a simple layout that is the same as the original circuit diagram. Figure 15.4 shows how the components could be placed. They must be positioned on the top side of the PCB and are drawn with thin lines; the tracks and pads on the bottom side of the PCB are drawn with wider lines. Compare this with Figure 15.3 and convince yourself that it is the same circuit. The transistor is assumed to be a BC184L – if you use a BC184 the leads will be in a different order. The variable resistor (R) in Figure 15.3 is drawn as a preset in Figure 15.4. A preset is a miniature

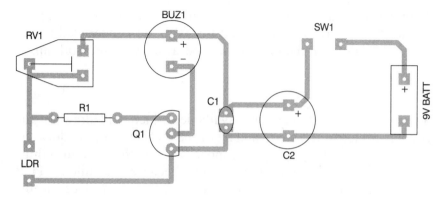

Figure 15.4

version of a potentiometer and is designed to be mounted directly on a PCB. A horizontally mounting mini-version is suggested (min hor); adjustments to the setting are made with a screwdriver. If a full-size potentiometer is preferred, i.e. one that can be fitted with a control knob, the connecting wires can be inserted into the pads intended for the preset.

Look at Figure 15.5. It is the same circuit, but now the pads and tracks have been arranged to allow the components to be packed closely together. It looks less like the circuit diagram, but that is a small price to pay. Do not make the PCB so small that it is difficult to build – be realistic in what can be achieved with ordinary constructing tools.

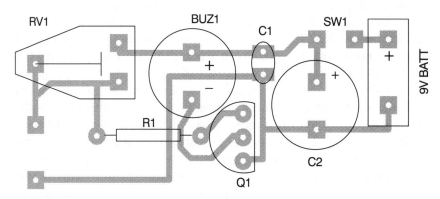

Figure 15.5

Processing a PCB

Now that you have a rough idea of your PCB layout, you will need to transfer the layout of pads and tracks to the copper-clad board. We will begin with the crudest method available, and work up.

Pen

Special 'etch resist' pens are available, which will draw on the copper surface of the board and resist the etchant. You can draw the tracks and pads directly on the copper, but note that you must draw the *mirror image* of your original layout, because you build the PCB with the copper side underneath. The board is then placed in the etchant, which dissolves the exposed copper but leaves intact the copper protected by the etch resist ink. The board can then be washed and dried and the ink rubbed off to leave the pattern of copper tracks required.

PCB transfers

Sets of transfers can be purchased, which are rubbed on to the copper board to create the layout required. These provide a much neater finish than the pen. The rest of the process is as before.

The two methods described cost the least but each PCB layout can only be used once. If a great deal of time is required to produce your layout it is better to employ a method which allows your design to be copied to more than one PCB.

Photo etch PCB

Photo etch PCBs are coated with a 'photo resist' material. The copper is coated with a layer of 'resist' which is sensitive to ultraviolet (UV) light. There is enough UV in sunlight to damage the coating, so a black plastic film is placed on this layer to protect it. The pen, or transfers, are used to produce the required layout on a transparent sheet, such as overhead projector film. There is no need to draw a mirror image – a correct image will be produced if the sensitized side of the board is placed against the layout. Don't forget to remove the black plastic material that protects the sensitized surface.

The photo-sensitive PCB is then exposed to UV light. Special UV boxes are available, but take care not to look at the UV lamps inside. Alternatively, the PCB could be exposed to sunlight for 5 or 10 minutes. An overhead projector provides enough UV light, given a sufficiently long exposure.

The PCB is then 'developed' in a weak solution of sodium hydroxide (or for best results use a PCB developer). This may be done in a developing tray, as used in photography, under ordinary room lighting. The image will have been copied to the PCB and the resist is dissolved wherever the sensitized surface was exposed to UV light. However, the tracks and pads remain on the copper. The PCB must now be etched in ferric chloride (or its equivalent) to remove all the areas of copper not protected by the remaining resist.

This method sounds complicated, but it is easier to do than to describe, and your PCB layout can be used to produce any number of boards.

Producing a PCB master

The master layout can be produced on overhead projector film as described, though a number of materials are available from electronics companies, which may provide better results. The material does not have to be transparent, as UV light will penetrate translucent materials, such as tracing paper. In fact, 'drafting paper' – a high quality form of tracing paper – works very well and costs much less than some of the alternatives.

Creating a PCB layout by means of transfers works well, but requires considerable patience, particularly if a change of design is required. Anyone using Electronics Workbench will have access to a computer and this is the ideal tool for producing a PCB master. Having created the design on the monitor, it can be printed out using suitable transparent or translucent film. Again, the use of drafting paper is recommended and works particularly well with laser printers.

Computer aided design

It is possible to use any artwork package to create a PCB layout, but anyone who has used a specialist PCB layout package will not wish to turn back. There are a number of packages available – almost all designed for the IBM system. Non-IBM users have a limited choice. IBM users can choose from very low cost 'basic' packages to all singing, all dancing wonders where the PCB virtually designs itself. A good quality system should allow you to paste and copy chunks of your design and move and rotate sections. All types of IC pin arrangements should be accommodated and the package should guide you through the process and have an effective 'help' system. The package should support double-sided PCBs (i.e. where tracks and pads can be placed on both sides of the board); more expensive systems will support multi-layer PCBs (unlikely to be required by non-industrial designers) and a 'silk layer', which enables the shapes and labels of components to be added. Industrial designs normally include a printed guide (the silk layer) on the component side of the PCB – while you are unlikely to require this, it is still helpful to produce a component layout printout.

Check whether the system operates only in DOS, or is designed for Windows. A windows package may be more 'friendly', particularly when shuffling files around the system, though it may require more memory. The more elaborate packages may require a fast computer – check the system requirements first. Most companies will provide an evaluation disk that demonstrates the system but lacks the ability to print and/or save. Many providers also offer a telephone hotline – quite important during the initial stages. Finally, check the price of the package and, if buying for a school or college, check the multi-user price and educational discount. A package that may seem expensive initially may be offered with attractive multi-user deals and, if the company then offers to provide all the students with their own non-printing versions free of charge, the deal can become very attractive!

Auto-routing

Auto-routing has been used for some time in industrial settings, but the increase in memory and operating speed of the home computer has made auto-routing a practical proposition. The system depends upon inputting the circuit diagram into the computer. A well-designed package will allow the circuit diagram to be drawn very quickly; the wiring will be similar to Workbench but the package will offer a much greater range of components and it will 'know' the layout of all standard ICs. For example the system should know that a CMOS 4001 NOR gate IC has a different pin arrangement to a CMOS 74HC02 NOR gate.

The system will then produce a 'netlist'. A netlist contains all the information regarding which points in a circuit should be connected together. The netlist is then transferred to the PCB layout package, which joins the appropriate points with tracks. The system will probably allow you the choice of where to place the components, but you will be guided by 'elastic strands' showing how the component you are placing is connected to the rest of the circuit. When all the components are placed the system will convert the elastic strands into tracks. The tracks will route themselves around other tracks and pads so as to avoid a short circuit. If a route is not possible, the package should report back to you. You may then reposition the component, or use a wire link to complete the connection.

Drilling the PCB

Returning to our layout, as shown in Figure 15.5, we will assume that the tracks and pads are now in the form of a copper pattern on the underside of your PCB. Each pad must now be drilled to allow the component leads to be inserted. A drill bit diameter of 1 mm will suit most component leads and small, inexpensive, PCB drills are available for the purpose. The preset may require slightly larger holes – check before building. Automatic drills are available – at a price – and this type of drill may be linked to the computer containing your PCB file. Given suitable software, the computer will control the movements of the drill – leaving you to do something more interesting.

Building your circuit

The components may now be positioned on the non-copper side of the PCB, with their leads pushed through the holes in the pads. Begin with the smallest components and work up in size. Some components, such as diodes, transistors and electrolytic (i.e. polarized) capacitors, must be fitted the correct way round. Solder all the leads securely to the copper pads and trim the ends to produce a neat result.

Testing

Test the circuit with care, ensuring that the battery is connected with the correct polarity. If a 100 mA, 5 V regulated supply is available, employ this for testing, as 100 mA is insufficient to destroy your circuit, no matter how many mistakes have been made. Otherwise, use a PP3 battery and check that components are not heating up. Adjust the setting on the potentiometer so that the buzzer is on the verge of sounding. Now shade the LDR. The buzzer should sound, indicating that all is well.

Fault finding

The most common causes of circuit failure are:

1 dry joints – where insufficient heat is used and solder does not bond with the copper pad or component lead
2 bridged pads, i.e. pads and/or tracks shorted together with solder
3 components inserted the wrong way round.

A voltmeter may be used to check the voltage at the transistor base – just like testing the transistor using the Workbench voltmeter. In fact, a number of readings may be taken around the circuit by connecting the negative side of the voltmeter to 0 V in the circuit and using the positive side of the voltmeter as a probe. Compare your results in practice with the results in Workbench, to locate the fault.

Further designs

We will now look at two more designs, both based around ICs.

Microphone amplifier

A microphone generates a very small electrical signal and will not work well if connected directly to the auxiliary input of a power amplifier, as the aux input requires a signal of around 1 V. What we require is a microphone amplifier that increases the amplitude of the signal by about 50 times without changing the frequency response. This means that a signal of 20 mV fed into our microphone amplifier will provide an output wave of amplitude 1 V.

We examined a variety of amplifier circuits in Chapter 5 and the most suitable for use as a microphone amplifier was the non-inverting amplifier. We will select a single rail version, as shown in Figure 5.14, as this may be powered from a single PP3 battery. The op-amp suggested for use in this circuit is a type 741 (e.g. LM741). Low noise versions of this IC are available for high quality work, making the finished project a very useful device.

Employing Figure 5.14 as the starting point, decoupling capacitors are now added, resulting in the circuit shown in Figure 15.6. The original circuit in Figure 5.14 would have worked perfectly well, but a number of other improvements have been made.

Resistor R1 has been added, so that if the microphone is disconnected there will still be a DC path from the input to ground. This will help reduce noise that may be present if the microphone is unplugged. The value of R1 is so high that it will have no other effect on the circuit. Capacitor C1 delivers the signal from the microphone to the circuit. C1 should have a large enough value to avoid attenuating (reducing) the bass frequencies. However, we

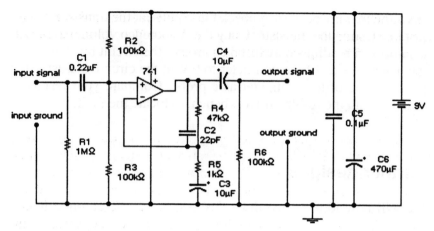

Figure 15.6

should avoid an electrolytic (polarized) capacitor, as these produce noise (unwanted hiss) – the last thing we want at the most sensitive side of the amplifier. Hence, C1 must be a non-polarized capacitor minimum value 0.22 μF (220 nF), but for the highest quality use a 0.47 μF (470 nF) polylayer type.

The gain of the circuit is set by the values of R4 and R5. The gain will be:

$$(R4/R5) + 1 = 48$$

As the output signal is larger than the input, capacitors C3 and C4 must have larger values than C1, forcing the use of electrolytic (polarized) types. However, the noise they cause is much less of a problem at the output side of the circuit. Capacitor C2 may be omitted, but its inclusion will help reduce any possibility of the circuit oscillating at high frequency. The value of C2 is 22 pF (22 picofarads); a 741 op-amp is very unlikely to oscillate, but the inclusion of C2 will do no harm.

A 9 V power supply is suggested, though the circuit will work happily on 12 V, or up to 30 V, providing the capacitors are rated at a sufficiently high voltage. It would be unwise to use a battery of less than 9 V, as our output signal may peak at over 2 V, i.e. 2 V above zero and 2 V below zero. This means that the total amplitude of the output signal may be over 4 V. The output of a 741 can only swing up and down by a volt or so less than the power supply, hence the maximum 'swing' on a 9 V supply is from 1 V to 8 V, i.e. a maximum amplitude of 7 V. If we used a 6 V battery, the output signal could be clipped, resulting in distortion.

Designing the layout

We have already seen how a circuit diagram is translated into a layout. When dealing with amplifiers (or any sensitive circuit), it is unwise to place output tracks and output components very close to input tracks and input components, as the output may induce interference into the input. The problem is especially acute if the gain of the amplifier is very high (e.g. if R4 is further increased in value).

When designing the PCB, we require the sizes of the components and the pin-out details of the op-amp. It is helpful to obtain all the components in advance; we will assume that the electrolytic capacitors are 'radial' types, i.e. their two leads emerge from the same end, so that they stand vertically on the PCB. Radial capacitors occupy less space than 'axial' types, which sit like resistors.

Figure 15.7 shows the pin layout of a 741 op-amp. Note that pins 1, 5 and 8 are not required in our circuit. However, they are connected internally and must be left unconnected in our layout.

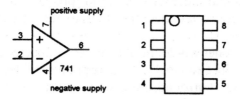

Figure 15.7

As the op-amp is the least flexible part of the circuit, i.e. its pin layout is fixed, we will position the op-amp first and arrange the other components around it. Figure 15.8 shows a suggested layout; we have kept the input and output sections of the circuit apart, yet have still produced a reasonably compact layout.

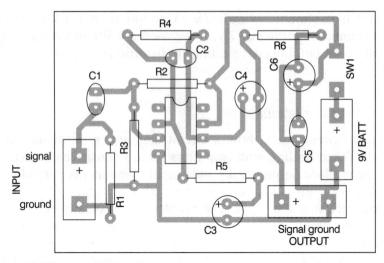

Figure 15.8

Constructing

The PCB should be produced as described earlier. It is unwise to solder an IC directly into place and an 8-pin DIL socket should be employed. Begin by soldering the socket into place, followed by the smallest components. Take care to fit the electrolytic capacitors the correct way round. Finally, insert the 741 op-amp into its socket, with the notch at the correct end. The IC is not sensitive to static electricity and no precautions are required. Solder the necessary connecting leads to the pads for the switch and power supply.

The microphone would normally be connected to the circuit via a jack plug and socket. Microphone cables consist of a protective screen surrounding a single wire. The screen should be connected to the 0 V or 'input ground' side of the circuit and the inner wire provides the 'input signal'. A 'balanced' microphone cable (often fitted to expensive microphones) contains two inner wires. In this case, connect one of the inner wires to the screen/ground, and use the other inner wire as the signal input. The balanced microphone will now operate as a normal type. If you wish to take full advantage of a balanced microphone, a differential microphone amplifier is required, as discussed in Chapter 6.

The output signal should be delivered to a power amplifier through the inner core of a screened cable. The screen should be connected to the 'output ground' pad on the microphone amplifier and the 'ground' side of the power amplifier.

Testing and fault finding

If a supply of 9 V–12 V, regulated at 100 mA, is available, this is ideal for testing, as it cannot damage your circuit, regardless of mistakes. Otherwise, test with a 9 V battery. Connect the output of the microphone amplifier to the input of a power amplifier, or to a pair of headphones. Connect a microphone to the input and switch on.

If fault finding is necessary, follow the fault finding section earlier in this chapter and use a voltmeter and/or oscilloscope to compare the actual values in your circuit with the predictions in Workbench. Note that pins 2, 3 and 6 of the IC should reside at a DC voltage equal to half that of the battery.

Choice of microphone

The microphone amplifier is designed for use with 'dynamic' microphones, or any other type that produces a signal without the need of an external power supply. Most cheaper 'electret' microphones contain a battery. Professional electret microphones generally require an external 'phantom power supply' of about 48 V. Clearly, no such provision has been made in this circuit. At the other end of the scale you can use an inexpensive electret 'insert' powered by connecting a 10 kΩ resistor between the positive power rail in your circuit and the input signal connection.

Egg timer

Our third example circuit is based on the egg timer originally shown in Figure 9.10 and developed for practical use as shown in Figure 15.9. As usual, we have added decoupling capacitors, the bulb has become an LED and it is controlled via a transistor to ensure that the logic level at the output of the NOR gate is not upset by drawing too much current. The buzzer is controlled via a transistor, for the same reason. The base resistor values of 4.7 kΩ are sufficiently high to limit the flow of current from each gate to a safe level, yet the high

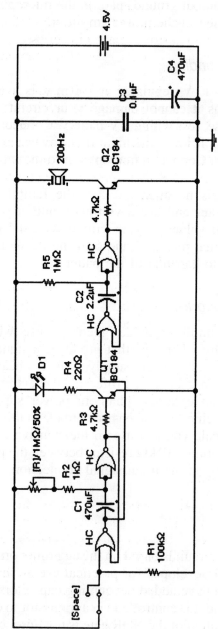

Figure 15.9

gain offered by the BC184 transistors ensures that sufficient current is available to operate the LED and buzzer.

A variable resistor has been added to the first timing circuit. The value of 1 MΩ provides a maximum time of: $1 \times 470 \times 0.7 = 329\,\text{s}$ (see Chapter 9). A 1 kΩ resistor is placed in series with the variable resistor to provide some resistance, even if the variable resistor is reduced to zero.

The second part of the circuit provides a signal to sound the buzzer at the end of the time period. The buzzer will sound for: $1 \times 2.2 \times 0.7 = 1.54\,\text{s}$. Note that there is no on/off switch; CMOS logic gates consume so little current that they may remain connected for many months without running down a battery. The switch shown in Figure 15.9 represents a momentary push-button switch, used to start the timer.

Simulating with workbench

The extra components will slow down the simulation and Workbench may be unable to 'reach a solution'. It will help if the decoupling capacitors are removed. Also, if the 'space' switch is closed *before* the circuit is activated, start-up may be more reliable. Workbench may complain that the polarized capacitors are connected the wrong way round – this is because they are reverse biased for a short time. It will be necessary to replace them with ordinary capacitors for the simulation.

Selecting components

We have chosen a 74HC series IC. Since this is a CMOS device, it will ensure that our battery will last a long time without having to include an on/off switch. However, 74HC ICs require a power supply of 2 V–6 V. Hence, we have chosen a 4.5 V battery (a 6 V battery might exceed 6 V when new). Our buzzer must be a solid state type capable of operating on 3 V–4 V. The transistors must be high-gain types, such as BC108, BC183 or BC184. We will select a BC184L, as the type tends to be more widely available. Note, however, that the L versions have the base lead on the *outside* of the three leads, rather than in the middle. Watch this when designing the PCB layout. The electrolytic (polarized) capacitors will occupy less space if they are radial types. The working voltage of all the capacitors must be at least 6 V. The variable resistor (potentiometer) is a linear type, so that the time period rises in equal steps as the shaft of the pot is rotated.

When creating the PCB layout, begin with the IC, as this is the most awkward device. Remember to connect up the power pins – pin 14 (top right-hand pin) should be connected to positive and pin 7 (bottom left) is the 0 V or negative supply pin. The four NOR gates are contained within a single IC known as type 74HC02. It is important to check the pin layout with a good quality catalogue or data book – note that type 74HC02 has a different pin-out to the 74HC08 mentioned in Chapter 8.

The potentiometer can be drawn in the form of a preset, so that you have a choice of using a pot linked via wires, or a preset fitted directly to the PCB. A pot is much easier to adjust, as it can be fitted with a control knob; a preset is less expensive, but must be adjusted with a screwdriver.

The suggested PCB layout is shown in Figure 15.10. The layout may not resemble the original circuit diagram, but if any two points on the circuit diagram are joined with a wire you will find that the same points are connected via a track in Figure 15.10. The components have been arranged with care to produce a reasonably

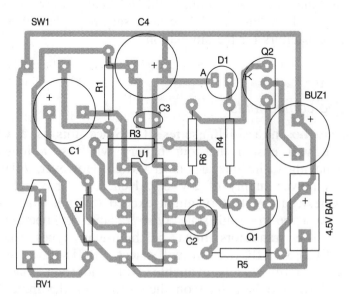

Figure 15.10

compact layout and to ensure that tracks can be routed successfully without crossing unrelated tracks. Sometimes it is impossible to route a track without crossing another; in this case, a pair of pads and a short wire link may be employed to 'jump over' the track. Each wire link will add to the cost of the PCB when manufactured by a company. Hence, use as few wire links as possible.

Constructing the PCB

As before, begin by inserting and soldering a 14 pin DIL socket for the IC. Next fit the smallest components, working up in size, checking the orientation of the electrolytic capacitors, LED (D1) and transistors. Connect the pot via wires, noting that the middle connection on the pot should be connected to the uppermost pad on the preset outline.

The buzzer and battery will probably be connected via leads; check which lead is which. The pads labelled SW1 are provided for the push-button switch used to trigger the timer.

Testing and fault finding

Test with a regulated 5 V supply as described earlier. When power is first applied, the monostables may be triggered, hence the LED may light and the buzzer may sound. Set RV1 (pot or preset) to its minimum resistance and wait for the circuit to stabilize. Then advance RV1 a little and press the push switch. The LED should light for the duration of the time period and the buzzer should sound for a short time at the end of the period.

If the circuit fails to work, follow the fault finding section earlier in this chapter and take voltage readings at key points in the circuit. Note that the current required by the voltmeter may upset the readings at the inputs to some of the gates. More reliable readings are possible at the outputs of gates.

It is easy to become depressed when a circuit fails to work immediately, but you will learn more by fault finding than at any other stage in the design process. Just one dry joint, one solder bridge, or one small error in the design will prevent the circuit working. It may be surprising that a circuit ever works – yet with persistence it will, and you will achieve a great deal of satisfaction. Good luck!

How to use the accompanying disk

The accompanying disk contains the sample circuits referred to in this book. These are arranged in folders by chapter within the folder called 'Samples'. Answers to the exercises set in this book are in a folder called 'Answers', also in the 'Samples' folder. You will also find a free demo version of Electronics Workbench, which allows you to view 15 of the sample circuits. Details of this demo are given in the file 'readme.txt' in the 'EWBdemo' folder.

If your CDROM drive has auto insert notification enabled, the demo will run once the CD is placed in the drive. (Pressing Escape on your keyboard will stop the program loading.) Otherwise, from Program Manager, select File menu and choose run. Type z:\RUNDEMO.EXE (where 'z' is your CD drive) and press ENTER.

To use the answer files and any of the samples not enabled in the demo you will need a copy of Electronics Workbench version 5.1 or higher.

The demo gives ordering information for obtaining a full copy of Electronics Workbench.

Index

This book is dedicated to
my parents and to
Joy Phillips

29089

Electronics Projects using Electronics Workbench